Classics in Mathematics

André Weil   Elliptic Functions according to Eisenstein and Kronecker

Springer
Berlin
Heidelberg
New York
Barcelona
Hong Kong
London
Milan
Paris
Singapore
Tokyo

André Weil

Elliptic Functions according to Eisenstein and Kronecker

Reprint of the 1976 Edition

Originally published as Vol. 88 of the
Ergebnisse der Mathematik und ihrer Grenzgebiete

Mathematics Subject Classification (1991): 11G

Library of Congress Cataloging-in-Publication Data

Weil, André, 1906-
Elliptic functions according to Eisenstein and Kronecker / André Weil.
p. cm. -- (Classics in mathematics, ISSN 1431-0821)
Originally published: Berlin; New York: Springer-Verlag, 1976,
in series: Ergebnisse der Mathematik und ihrer Grenzgebiete; 88.
Includes bibliographical references and index.
ISBN 3-540-65036-9 (soft : alk. paper)
1. Elliptic functions. I. Title. II. Series.
QA343.W45 1999
515'.983--dc21

Photograph of André Weil by kind permission of The Inamori Foundation, Kyoto

ISSN 1431-0821
ISBN 3-540-65036-9 Springer-Verlag Berlin Heidelberg New York

This work is subject to copyright. All rights are reserved, whether the whole or part of the material is concerned, specifically the rights of translation, reprinting, reuse of illustrations, recitation, broadcasting, reproduction on microfilm or in any other way, and storage in data banks. Duplication of this publication or parts thereof is permitted only under the provisions of the German Copyright Law of September 9, 1965, in its current version, and permission for use must always be obtained from Springer-Verlag. Violations are liable for prosecution under the German Copyright Law.

© Springer-Verlag Berlin Heidelberg 1999
Printed in Germany

The use of general descriptive names, registered names, trademarks etc. in this publication does not imply, even in the absence of a specific statement, that such names are exempt from the relevant protective laws and regulations and therefore free for general use.

SPIN 10684597   41/3143-5 4 3 2 1 0 – Printed on acid-free paper

André Weil

Elliptic Functions according to Eisenstein and Kronecker

Springer-Verlag
Berlin Heidelberg New York 1976

Prof. André Weil

The Institute for Advanced Study, School of Mathematics, Princeton, N.J. 08540 (USA)

AMS Subject Classification (1970): 33 A 25

ISBN 3-540-07422-8 Springer-Verlag Berlin Heidelberg New York
ISBN 0-387-07422-8 Springer-Verlag New York Heidelberg Berlin

Library of Congress Cataloging in Publication Data. Weil, André, 1906 –. Elliptic functions according to Eisenstein & Kronecker. (Ergebnisse der Mathematik und ihrer Grenzgebiete; 88). Bibliography: p. Includes index. 1. Functions, Elliptic. I. Title. II. Series. QA343.W45   515'.983   75-23200

This work is subject to copyright. All rights are reserved, whether the whole or part of the material is concerned, specifically those of translation, reprinting, re-use of illustrations, broadcasting, reproduction by photocopying machine or similar means, and storage in data banks. Under § 54 of the German Copyright Law where copies are made for other than private use, a fee is payable to the publisher, the amount of the fee to be determined by agreement with the publisher.
© by Springer-Verlag Berlin Heidelberg 1976.
Printed in Germany.
Monophoto typesetting and offset printing: Zechnersche Buchdruckerei, Speyer.
Bookbinding: Konrad Triltsch, Würzburg.

Editorial Preface

It is a great pleasure to my colleagues on the editorial board of the series Ergebnisse der Mathematik and myself to welcome this work, *Elliptic Functions according to Eisenstein and Kronecker*, by André Weil. However, some readers may be surprised to find in this series a text which appears at first sight to be in essence a contribution to the history of mathematics, and which would therefore seem to be very untypical of the books in this series. However, the editors are strongly of the opinion that, while this text undoubtedly contributes notably to the history of our science, it is also of great value to contemporary mathematical research. Thus we had no hesitation in asking the permission of Professor Weil to publish his text in our series; and we are delighted to have had his agreement to do so.

PETER HILTON
Chairman, Editorial Board, Ergebnisse der Mathematik

Battelle Seattle Research Center, August, 1975

Foreword

"When kings are building", says the German poet, "carters have work to do". Kronecker quoted this, in his letter to Cantor of September 1891, only to add, thinking of himself no doubt, that each mathematician has to be king and carter at the same time.

But carters need roads. Not seldom, in the history of our science, has it happened that a king opened up a new road into the promised land and that his successors, intent upon their own paths, allowed it to be overrun by brambles and become unfit for transit.

To help clean up such a road is the purpose of this little book, arising out of lectures given at the Institute for Advanced Study in the Fall of 1974. I am grateful to Melvyn Nathanson for having put at my disposal his notes from those lectures. Where the road will lead remains in large part to be seen, but indications are not lacking that fertile country lies ahead.

On the other hand, since much of the material in this volume seems suitable for inclusion in elementary courses, it may not be superfluous to point out that it is almost entirely self-contained. Even the basic facts about trigonometric functions are treated *ab initio* in Chapter II, according to Eisenstein's method. It would have been both logical and convenient to treat the gamma-function similarly in Chapter VII; for the sake of brevity, this has not been done, and a knowledge of some elementary properties of $\Gamma(s)$ has been assumed. One further prerequisite in Part II is Dirichlet's theorem on Fourier series, together with the method of Poisson summation which is only a special case of that theorem; in the case under consideration (essentially no more than the transformation formula for the theta-function) this presupposes the calculation of some classical integrals. Distributions occur only in § 10 of Chapter VII and §§ 16—18 of Chapter VIII; as these are disjoint from the rest of the book, they can be skipped without damage, if perhaps not without loss. In Chapter VIII, the Bessel function K_ν occurs inevitably, as a definite integral, and I have of course adopted for it the usual notation; no property of that function, or rather of that integral, is used beyond the most obvious ones. As to the final chapter, it concerns applications to number-theory, and there, of course, some number-theory is needed.

Princeton, the 21st of March 1975.   ANDRÉ WEIL

Table of Contents

Part I EISENSTEIN

Chapter I
Introduction 3

Chapter II
Trigonometric functions 6

Chapter III
The basic elliptic functions 14

Chapter IV
Basic relations and infinite products 22

Chapter V
Variation I 35

Chapter VI
Variation II 42

Part II KRONECKER

Chapter VII
Prelude to Kronecker 51

Chapter VIII
Kronecker's double series 69

Chapter IX
Finale: Allegro con brio
(Pell's equation and the Chowla-Selberg formula) 87

Index of Notations 93

Part I

EISENSTEIN

Chapter I

Introduction

In 1891, Kronecker agreed to give the inaugural lecture at the first meeting of the newly founded German Mathematical Society. He cancelled this plan after losing his wife, but (in a letter to Cantor, president of the society) expressed the hope that he would still be able to supply a written text for the lecture, which he described as follows:

„*Der Vortrag... sollte kurzweg den Titel haben „Über Eisenstein"... Dabei müßten dann außer den rein arithmetischen und analytisch-arithmetischen noch ganz besonders seine rein analytischen Untersuchungen über elliptische Funktionen hervorgehoben werden, welche dem Bewußtsein der Jetztzeit ganz abhanden gekommen sind...*"[1]

Soon after that, Kronecker died; he never wrote up that lecture. However, he had already discussed Eisenstein's work at some length (pointing out how Eisenstein had anticipated some of Weierstrass' best-known innovations and gone well beyond them) in his last major paper on elliptic functions, printed in 1891 by the Berlin Academy. This is how he comments upon it:

"Essentially new points of view... particularly concerning the transformation theory of theta-functions... were introduced by Eisenstein in the fundamental but seldom quoted *Beiträge zur Theorie der elliptischen Functionen* published in *Crelles Journal* in 1847, which are based upon entirely original ideas..."

When Kronecker expresses himself with such enthusiasm, it is rather obvious that he has only just re-discovered the paper in question; he goes on to point out its relevance to his current research, which he had clearly not noticed until that moment[2]. In both of the above quotations, the reference is to Eisenstein's paper *Genaue Untersuchung der unendlichen Doppelproducte, aus welchen die elliptischen Functionen als Quotienten zusammengesetzt sind, und der mit*

---

[1] "The lecture was to be entitled simply "On Eisenstein". Its theme was to be, not so much his work on number-theory or those of his papers which combine number-theory with function-theory, but more specifically, and with particular emphasis, his purely analytical investigations on elliptic functions, which have been so profoundly forgotten that they now seem to be as good as lost." (Kronecker, *Werke*, vol. V, p. 499.)

[2] Kronecker, *Werke*, vol. V, p. 149.

ihnen zusammenhangenden Doppelreihen; this is part VI of his *Beiträge zur Theorie der elliptischen Functionen*; it was published in vol. 35 (1847) of *Crelles Journal*, pp. 153—274, and reproduced in Eisenstein's volume *Mathematische Abhandlungen* published in 1847 with a preface by Gauss.

Well could Kronecker say of that paper that it was "seldom quoted"; it is doubtful whether there is a single reference to it, apart from Kronecker's and from a footnote in Hurwitz' thesis[3], in all the mathematical literature of the XIXth century; in the present century one could perhaps find two or three more. Eisenstein's ideas could indeed seem "as good as lost".

It is not merely out of an antiquarian interest that the attempt will be made here to resurrect them. Not only do they provide the best introduction to much of the work of Hecke; but we hope to show that they can be applied quite profitably to some current problems, particularly if they are used in conjunction with Kronecker's late work which is their natural continuation. Perhaps their range of usefulness can even be extended beyond the theory of elliptic functions and of the modular group, and in particular to the arithmetical study of the Eisenstein series for the Hilbert modular group[4]; but this will not be discussed here.

As any reader of Eisenstein must realize, he felt hard pressed for time during the whole of his short mathematical career. As a young man he complains of nervous ailments which often compel him to interrupt his work; later, he developed tuberculosis, and died of it in 1852 at the age of 29. His papers, although brilliantly conceived, must have been written by fits and starts, with the details worked out only as the occasion arose; sometimes a development is cut short, only to be taken up again at a later stage. Occasionally Crelle let him send part of a paper to the press before the whole was finished. One is frequently reminded of Galois' tragic remark "Je n'ai pas le temps".

In view of this, it would be foolish to follow Eisenstein step by step; we shall feel free to re-arrange his material as he might have done himself on more mature consideration, and to make use of his own hints in order to improve upon his exposition when this does no violence to his way of thinking.

One general observation is needed concerning questions of convergence. In Eisenstein's days, the concept of absolute convergence (as opposed to "conditional" convergence) was still comparatively new; he learnt it, he says, from Dirichlet's paper on the arithmetic progression, and he is quite careful in using it whenever needed. For instance, the earlier part of the paper to be studied here is devoted to a proof for the convergence of the series

$$\sum (n_1^2 + n_2^2 + \cdots + n_\nu^2)^{-\sigma}$$

---

[3] Hurwitz, *Werke*, vol. I, p. 31.
[4] This prediction has been fulfilled (even beyond my own expectations) since the above lines were written; cf. G. Shimura, *On some arithmetic properties of modular forms of one and several variables*, to appear in *Ann. of Math.*

with $\sigma > \nu/2$, and of more general series of the same kind. Nowadays such results are a matter of common knowledge and may be taken for granted. On the other hand, Eisenstein is unaware of the concept of uniform convergence; he assumes tacitly, and without proof, that the series of analytic functions which he introduces can be differentiated term by term; perhaps this was why Weierstrass ignored his work. Actually the gap is easily filled; if challenged to do so, Eisenstein might well have proceeded as follows. Take as a typical example the case of the series $\sum (x+\mu)^{-n}$ which occur in his theory of the trigonometric functions (cf. Chap. II). Discarding a finite number of terms, we have to consider the absolutely convergent series

$$f_n(x) = \sum_{\mu=M}^{+\infty} (x+\mu)^{-n} + \sum_{\mu=M}^{+\infty} (x-\mu)^{-n} \quad (n \geq 2)$$

$$f_1(x) = \sum_{\mu=M}^{+\infty} \left(\frac{1}{x+\mu} + \frac{1}{x-\mu} \right),$$

where M is an integer >1. Call $f(x)$ any one of these series; write $\varphi_\mu(x)$ for its μ-th term, and expand $\varphi_\mu(x+y)$, by the binomial formula, into a power-series in y:

$$\varphi_\mu(x+y) = \sum_{m=0}^{+\infty} \varphi_{\mu,m}(x) y^m.$$

A trivial estimate will then show that the double series $\sum_{\mu,m} \varphi_{\mu,m}(x) y^m$ is absolutely convergent for $|x| \leq M-1$, $|y| < 1$. Consequently we can write

$$f(x+y) = \sum_{m=0}^{+\infty} \left(\sum_\mu \varphi_{\mu,m}(x) \right) y^m.$$

As the coefficients of the y^m are nothing else, up to obvious constant factors, than the series derived from the series for $f(x)$ by successive differentiation term by term, this guarantees the legitimacy of Eisenstein's procedure. In what follows, such matters will be taken for granted once and for all.

Chapter II

Trigonometric Functions

§ 1. As Eisenstein shows, his method for constructing elliptic functions applies beautifully to the simpler case of the trigonometric functions. Moreover, this case provides, not merely an illuminating introduction to his theory, but also the simplest proofs for a series of results, originally discovered by Euler, which will have to be used later on.

The method is based on the consideration of the series

$$\varepsilon_n(x) = \sum_{\mu=-\infty}^{+\infty} (x+\mu)^{-n},$$

where n is an integer ≥ 1. This requires no comment if $n>1$. In order to deal with the case $n=1$, we introduce a symbol $\sum_e$, to be called "Eisenstein summation" (for simple series), defined by

$$\sum_{\mu}{}_{e} = \lim_{M \to +\infty} \sum_{\mu=-M}^{+M}.$$

Then we define ε_1 by putting

$$\varepsilon_1(x) = \sum_{\mu}{}_{e} \frac{1}{x+\mu}.$$

As the series for ε_n is absolutely convergent for $n \geq 2$, it is obvious that, for such n, ε_n is periodic of period 1; the same is true of ε_1 because the terms of the series for ε_1 tend to 0 for $\mu \to \pm\infty$. Differentiating term by term (cf. the final remarks in Chap. I), we get $d\varepsilon_n/dx = -n\varepsilon_{n+1}$ for all $n \geq 1$. Expanding $(x+\mu)^{-n}$, for $n \geq 1$, $\mu \neq 0$, $|x| < 1$, into a power-series in x, we find for $\varepsilon_n(x) - x^{-n}$ a power-series in x, convergent for $|x|<1$. For $n=1$, we will write this as

$$\varepsilon_1(x) = \frac{1}{x} - \sum_{m=1}^{+\infty} \gamma_m x^{m-1}$$

where the coefficients γ_m are 0 when m is odd, and otherwise are given by

$$\gamma_{2m} = 2 \sum_{\mu=1}^{+\infty} \mu^{-2m}.$$

Differentiating $n-1$ times, we get:

$$\varepsilon_n(x) = \frac{1}{x^n} + (-1)^n \sum_{m=1}^{+\infty} \binom{2m-1}{n-1} \gamma_{2m} x^{2m-n},$$

where the "binomial coefficients" $\binom{2m-1}{n-1}$ are 0 for $2m<n$. Clearly $\varepsilon_n(x)$ is an even or an odd function of x according as n is even or odd. For each $n \geq 1$, γ_n is the value of $\varepsilon_n(x) - x^{-n}$ at $x=0$.

§ 2. The question is now to construct non-linear identities between the functions ε_n; the starting point for this is supplied, according to Eisenstein, by identities for rational functions. Take two independent variables p, q, and put $r = p+q$. Dividing by pqr, we get

(1) $$\frac{1}{pq} = \frac{1}{pr} + \frac{1}{qr}.$$

More generally, if m, n are two integers ≥ 1, we have

(2) $$\frac{1}{p^m q^n} = \sum_{h=0}^{m-1} \frac{n(n+1)\ldots(n+h-1)}{h! \, p^{m-h} r^{n+h}} + \sum_{k=0}^{n-1} \frac{m(m+1)\ldots(m+k-1)}{k! \, q^{n-k} r^{m+k}}.$$

This can be derived from (1) by successive differentiation, $m-1$ times with respect to p and $n-1$ times with respect to q. Alternatively, (2) may be regarded as the partial fraction decomposition of $p^{-m}(r-p)^{-n}$ regarded as a rational function of p when r is taken as a constant. For $m=n=2$, we get:

(3) $$\frac{1}{p^2 q^2} = \frac{1}{p^2 r^2} + \frac{1}{q^2 r^2} + \frac{2}{pr^3} + \frac{2}{qr^3}.$$

In (3), put $p = x+\mu$, $q = y+v-\mu$; also, put $z = x+y$; then $r = z+v$. Apply now "Eisenstein summation" to (3) with respect to μ, while v is kept constant. This gives:

$$\sum_{\mu}{}_e (p^{-2} q^{-2} - p^{-2} r^{-2} - q^{-2} r^{-2}) = 2r^{-3} [\varepsilon_1(x) + \varepsilon_1(y+v)],$$

where we may replace $\varepsilon_1(y+v)$ by $\varepsilon_1(y)$, and $\sum_e$ by $\sum$ since the series is absolutely convergent. Now summation with respect to v gives

(4) $$\varepsilon_2(x)\varepsilon_2(y) - \varepsilon_2(x)\varepsilon_2(z) - \varepsilon_2(y)\varepsilon_2(z) = 2\varepsilon_3(z)[\varepsilon_1(x) + \varepsilon_1(y)]$$

since everything now is absolutely convergent. This may be regarded as an addition formula for the ε-functions.

§ 3. For a given x, not an integer, both sides of (4), regarded as functions of y, have a double pole at $y=0$; expanding them into power-series in y, one verifies at once that the terms in y^{-2} and y^{-1} have the same coefficients on both sides. Equating the constant terms, we get

(5) $$3\varepsilon_4(x) = \varepsilon_2(x)^2 + 2\varepsilon_1(x)\varepsilon_3(x).$$

Similarly, for a given x, regard both sides of (4) as functions of z and expand them at $z=0$; equating the constant terms, we get

(6) $$\varepsilon_2(x)^2 = \varepsilon_4(x) + 2\gamma_2\varepsilon_2(x).$$

This gives $\varepsilon_1\varepsilon_3 = \varepsilon_2^2 - 3\gamma_2\varepsilon_2$; differentiating this, we get $\varepsilon_2\varepsilon_3 - 2\gamma_2\varepsilon_3 = \varepsilon_1\varepsilon_4$; combined with (6), this gives $\varepsilon_3 = \varepsilon_1\varepsilon_2$. Substituting $\varepsilon_1\varepsilon_2$ for ε_3 in the formula for $\varepsilon_1\varepsilon_3$ and dividing by ε_2, we get $\varepsilon_1^2 = \varepsilon_2 - 3\gamma_2$. As we have $\varepsilon_2 = -d\varepsilon_1/dx$, this shows that ε_1 is the solution of the differential equation $dX/dx = -X^2 - 3\gamma_2$ which becomes infinite at $x=0$. It is well known, of course, that this implies $\varepsilon_1(x) = \pi \cot \pi x$ (we write cot for the cotangent).

§ 4. It is more interesting, however, to proceed as if one had no previous knowledge of the trigonometric functions, and to regard the differential equation in question as defining the cotangent. More precisely, put $a = (3\gamma_2)^{-\frac{1}{2}}$, $a > 0$. From the above construction, it follows that $u \to a\varepsilon_1(au)$ is a solution of the differential equation $dv/du = -u^2 - 1$, with the period a^{-1}; it is clear that any two solutions of that equation can differ only by a translation on u; therefore this equation has a unique solution which becomes infinite for $u = 0$. If we *define* this solution to be $v = \cot u$, and if we *define* π to be its period, we can now write $\varepsilon_1(x) = \pi \cot \pi x$, and we have $\gamma_2 = \pi^2/3$.

From this point on, one can develop in various manners the elementary theory of the trigonometric functions; this can be done either by making use of the formulas obtained above or by deriving further identities for trigonometric functions from identities for rational functions according to the method described in § 2. As an example, take the addition formula for the cotangent:

(7) $$2\varepsilon_1(x+y)[\varepsilon_1(x) + \varepsilon_1(y)] = [\varepsilon_1(x) + \varepsilon_1(y)]^2 - \varepsilon_2(x) - \varepsilon_2(y).$$

Eisenstein proves this as follows. First observe that, for every integer v, we have

(8) $$\varepsilon_1(x)+\varepsilon_1(y) = \sum_\mu \left(\frac{1}{x+\mu} + \frac{1}{y+v-\mu}\right),$$

the series being absolutely convergent. Put now

$$z=x+y, \quad p=x+\mu, \quad q=y-\mu, \quad p'=x+\mu-v, \quad q'=y+v-\mu,$$

and apply (1) to p,q' and to p',q. We get

$$\frac{1}{pq'}+\frac{1}{p'q} = \frac{1}{p+q'}\left(\frac{1}{p}+\frac{1}{q'}\right) + \frac{1}{p'+q}\left(\frac{1}{p'}+\frac{1}{q}\right),$$

which may be regarded as the partial fraction decomposition of the left-hand side when it is regarded as a function of x alone, z, μ and v being kept constant. Write A for the right-hand side. Similarly, applying (1) to $p, -p'$ and then to $q, -q'$, we get, for $v \neq 0$:

$$\frac{1}{pp'}+\frac{1}{qq'} = \frac{1}{v}\left(\frac{1}{p'}-\frac{1}{p}+\frac{1}{q}-\frac{1}{q'}\right);$$

write B_v for the right-hand side, and put $B_0 = p^{-2}+q^{-2}$. This gives the identity

$$\left(\frac{1}{p}+\frac{1}{q}\right)\left(\frac{1}{p'}+\frac{1}{q'}\right) = A+B_v.$$

Sum this with respect to μ, v being kept constant; then perform Eisenstein summation, i.e. $\sum_e$, on v; in the first summation, everything is absolutely convergent. Moreover, in the left-hand side, even double summation, on (μ, v), would be absolutely convergent; therefore, in the left-hand side, one would get the same result by performing summation on μ and on $\mu - v$ independently, which, in view of (8), gives

$$[\varepsilon_1(x)+\varepsilon_1(y)]^2.$$

As to the summation on A, it gives the left-hand side of (7); the summation on B_0 gives the last two terms in (7), and the summation on B_v for $v \neq 0$ gives 0. This proves (7).

Alternatively, consider the formula

(9) $$\varepsilon_1(z-x)[\varepsilon_2(x)-\varepsilon_2(z)]-\varepsilon_1(x)\varepsilon_2(z)-\varepsilon_1(z)\varepsilon_2(x)=0,$$

which, in view of the identity $\varepsilon_2 = \varepsilon_1^2 + \pi^2$ of §3, is trivially equivalent to (7). Put $y = z - x$, and write $f(x, y)$ for the left-hand side of (9); then (4) gives $\partial f/\partial y = 0$, so that, instead of $f(x, y)$, we may write $f(x)$. As this left-hand side is symmetric in x and z, we have $f(x) = f(z)$ for all x, z, so that f is constant; changing x, z into $-x, -z$, the left-hand side changes sign, so that f is odd. Therefore $f = 0$; this again proves (7).

§5. We merely note here that §3 of Eisenstein's paper contains brief indications about far more general identities for trigonometric functions which can be deduced by his method from the corresponding identities for rational functions and in their turn (as we shall see) can be used in order to derive identities for elliptic functions. We also note his incidental mention of the formula

$$\frac{1}{\sin u} = \frac{1}{2} \cot \frac{u}{2} - \frac{1}{2} \cot \frac{u+\pi}{2},$$

which can be rewritten as

(10) $$\frac{\pi}{\sin \pi x} = \frac{1}{2} \varepsilon_1 \left(\frac{x}{2}\right) - \frac{1}{2} \varepsilon_1 \left(\frac{1+x}{2}\right) = \sum_e \frac{(-1)^v}{x+v}.$$

This may be regarded as *defining* the sine function. Now (7) gives the multiplication formula

$$2\varepsilon_1(2x)\varepsilon_1(x) = 2\varepsilon_1(x)^2 - \varepsilon_2(x) = \varepsilon_1(x)^2 - \pi^2.$$

Here substitute either $\frac{x}{2}$ or $\frac{1+x}{2}$ for x; this shows that $\varepsilon_1\left(\frac{x}{2}\right)$ and $\varepsilon_1\left(\frac{1+x}{2}\right)$ are the roots of the equation

$$Y^2 - 2\varepsilon_1(x) Y - \pi^2 = 0,$$

and (10) gives

$$\pi/\sin \pi x = \varepsilon_2(x)^{\frac{1}{2}}, \qquad \varepsilon_2(x) = (\pi/\sin \pi x)^2.$$

Differentiating this, and using the formulas of §3, one gets

(11) $$\varepsilon_1(x) = \frac{d}{dx} \log \sin \pi x.$$

Had Eisenstein lived longer and pursued his investigations further, he might well have been led by (10) and other formulas of the same kind to the consideration of the more general series of the form $\sum \chi(v)(x+v)^{-n}$, where χ is a character (not necessarily of finite order) of the additive group of integers, and of the corresponding series in the theory of elliptic functions. As we shall see, this topic was eventually taken up by Kronecker (cf. Chap. VII and VIII).

§ 6. Now we introduce infinite products; this calls for some preliminary remarks. Let $P = \prod p_\mu$ be such a product, where some of the p_μ may be 0; we extend to it all our definitions and remarks concerning series by taking logarithms, with the understanding that finitely many factors should always be disregarded when this is appropriate. This implies that the product is not called convergent unless p_μ tends to 1 for $\mu \to \pm \infty$ (so that only finitely many factors can be 0); then, for $\log p_\mu$, one always understands the principal branch of the log in some neighborhood of 1; outside that neighborhood, one may take for $\log p_\mu$ any branch of the log (e.g., for definiteness, the one which is of the form $\log|p_\mu| + \pi i t$ with $-1 \leq t < 1$), with the understanding that $\log 0 = \infty$; that being so, we write

$$\log P = \sum \log p_\mu$$

when the product converges; the log in the left-hand side may then be any branch of the log function. The product is said to be absolutely convergent if the series $\sum \log p_\mu$ is so (here one has to disregard the finitely many factors $p_\mu = 0$). The symbol $\prod_e$ has to be understood accordingly; in view of our definition of $\sum_e$, it may be regarded as defined by the formula

$$\prod_\mu{}_e p_\mu = \lim_{M \to +\infty} \prod_{\mu=-M}^{+M} p_\mu = \prod_{\mu=-M}^{+M} p_\mu \cdot \prod_{\mu=M+1}^{+\infty} (p_\mu p_{-\mu}),$$

this being meaningful only if $p_\mu p_{-\mu}$ tends to 1 for $\mu \to +\infty$ (while p_μ need not do so).

It is now meaningful to consider the product

$$P(x) = \prod_\mu{}'_e \left(1 + \frac{x}{\mu}\right),$$

where $\prod'$, as usual, means the product taken over all $\mu \neq 0$; moreover, we may write

$$\log P(x) = \sum_\mu{}'_e \log\left(1 + \frac{x}{\mu}\right).$$

In view of the remarks at the end of Chap. I, we may differentiate term by term; at this point, however, Eisenstein feels the need for some justification and proceeds, as one might do nowadays, by differentiating formally and then integrating the formula obtained in this manner; he fails to notice that even this falls short of what is required, in the absence of any reference to what we now call uniform convergence. Anyway, this gives

$$\frac{d}{dx}\log P(x) = \sum_e' \frac{1}{x+\mu} = \varepsilon_1(x) - \frac{1}{x}.$$

In view of (11), it shows that $xP(x)/\sin \pi x$ is a constant; (10) shows that it has the value $1/\pi$ for $x=0$. This gives the classical Euler product for the sine function in the form

(12) $$\sin \pi x = \pi x \prod_e' \left(1 + \frac{x}{\mu}\right).$$

From this one deduces easily the following formula, valid whenever x/u is not an integer:

(13) $$\frac{\sin \pi \frac{x-t}{u}}{\sin \pi \frac{x}{u}} = \prod_e \left(1 - \frac{t}{x+\mu u}\right);$$

this will be useful further on.

§ 7. So far it has not been necessary to say whether we were dealing with real or with complex numbers. Everything is valid in both cases, except for the proof of (12), which, for real variables, would only show that the ratio of both sides is 1 in the interval $[-1,1]$ and constant in each interval $[\mu,\mu+1]$; to complete it, one should verify (as can easily be done) that the right-hand side is periodic of period 2.

In the complex case, one may also wish to derive from the above considerations the relation between trigonometric and exponential functions. This is done by Eisenstein by rewriting the addition formula (7) in the equivalent form

$$\varepsilon_1(x+y) = \frac{\varepsilon_1(x)\varepsilon_1(y) - \pi^2}{\varepsilon_1(x) + \varepsilon_1(y)}.$$

Now introduce the function

$$\mathbf{e}(x) = \frac{\varepsilon_1(x) + \pi i}{\varepsilon_1(x) - \pi i};$$

its addition formula is at once seen to be

$$\mathbf{e}(x+y) = \mathbf{e}(x)\mathbf{e}(y)$$

and its expansion at $x=0$ begins with $1+2\pi i x$. Therefore we have $\mathbf{e}(x)=e^{2\pi i x}$, and we get the usual formulas, at first for the cotangent and then for the sine and cosine. For future use, we note the formula:

(14) $$\varepsilon_1(x) = \pi i \frac{\mathbf{e}(x)+1}{\mathbf{e}(x)-1}.$$

Defining the Bernoulli numbers B_m, as usual, by means of the power-series

$$\frac{1}{2}\frac{e^v+1}{e^v-1} = \frac{1}{v} - \sum_{m=1}^{\infty} (-1)^m B_m \frac{v^{2m-1}}{(2m)!},$$

we get now the power-series for $\varepsilon_1(x)$ in the form

(15) $$\varepsilon_1(x) = \frac{1}{x} - \sum_{m=1}^{\infty} (2\pi)^{2m} B_m \frac{x^{2m-1}}{(2m)!}.$$

In view of the formulas in § 1, this gives

(16) $$\gamma_{2m} = 2\sum_{\mu=1}^{\infty} \mu^{-2m} = (2\pi)^{2m} \frac{B_m}{(2m)!}.$$

Chapter III

The Basic Elliptic Functions

§ 1. From now on, we deal with complex variables. We write W for a lattice in the complex plane; we write u, v for two generators of W, so that W consists of the points $w = \mu u + v v$, where μ, v are integers. Then v/u is not real and may be written as $v/u = \delta \tau$, where $\delta = \pm 1$ and τ is in the upper half-plane; sometimes it will be convenient to write $\delta(u,v)$ for δ. We write $q = \mathbf{e}(\tau)$, with $\mathbf{e}$ as defined in Chap. II, § 7, and we will always take for $\sqrt{q} = q^{\frac{1}{2}}$ the branch given by $q^{\frac{1}{2}} = \mathbf{e}(\tau/2)$; we have $|q| < 1$. Since $u\bar{v} - \bar{u}v$ has the value $\delta u \bar{u}(\bar{\tau} - \tau)$, we can write it in the form

$$u\bar{v} - \bar{u}v = -2\pi i \delta A, \quad A > 0.$$

§ 2. Now we introduce the series

$$E_n(x) = \sum_{w \in W} (x+w)^{-n}.$$

This requires no comment for $n \geq 3$, since then the series are absolutely convergent; moreover, in view of the remarks at the end of Chap. I, we have $dE_n/dx = -n E_{n+1}$.

For $n=1$ and $n=2$, Eisenstein makes use of a summatory process which we shall call *Eisenstein summation*; it depends upon the choice of the generators u, v for W and is given by the formula

$$\sum_e = \sum_e \left(\sum_e \right) = \lim_{N \to \infty} \sum_{v=-N}^{N} \left(\lim_{M \to \infty} \sum_{\mu=-M}^{M} \right),$$

where we have put $w = \mu u + v v$. Clearly the summation with respect to μ can be performed explicitly by means of the formulas in Chap. II; in fact, we have, for all $n \geq 1$:

(1) $$\sum_{\mu} {}_e (x+w)^{-n} = u^{-n} \varepsilon_n \left(\frac{x+vv}{u} \right),$$

and therefore:

(2) $$\sum_e (x+w)^{-n} = u^{-n} \varepsilon_n\left(\frac{x}{u}\right) + u^{-n} \sum_1^\infty \left(\varepsilon_n\left(\frac{x+v v}{u}\right) + \varepsilon_n\left(\frac{x-v v}{u}\right)\right).$$

In §6 we shall give for this a more explicit formula, which will make it clear that the series in (2) is absolutely convergent, and (in view of the remarks at the end of Chap. I) that it can be differentiated term by term. Taking this for granted for the moment, we now define E_n, for all $n \geq 1$, by the formula

(3) $$E_n(x) = \sum_e (x+w)^{-n} = \sum_e (x + \mu u + v v)^{-n};$$

for this (just as for $\varepsilon_n(x)$ in his treatment of trigonometric functions) Eisenstein uses the notation (n, x). As it depends, not only upon x and the lattice W, but also upon the choice of the generators u, v for W, we shall write for it, more explicitly, $E_n(x; u, v)$ whenever necessary; for $n \geq 3$, this may also be written as $E_n(x; W)$. In view of the definition of $\sum_e$, $E_n(x)$ is an even or an odd function of x according as n is even or odd; in particular, we have $E_1(-x) = -E_1(x)$.

As termwise differentiation is allowed on (3), we have, for all $n \geq 1$, $dE_n/dx = -n E_{n+1}$.

§ 3. In the summation process $\sum_e$, one could interchange the roles of μ and v; this amounts to interchanging the generators u and v for W. More generally, any change of generators for W would produce a modified summation process. One may also make a translation on the lattice W, substituting, say, $(\mu' + \mu_0, v' + v_0)$ for (μ, v), and then applying $\sum_e$ to μ', v' instead of applying it to μ, v; this amounts to the same as substituting $x + w_0$ for x in $E_n(x)$, with $w_0 = \mu_0 u + v_0 v$. Finally, we may take a sublattice W' of W, with the generators u', v', and a set R of representatives for W/W' in W, and apply to any function $f(w)$ on the lattice W, e.g. to $f(w) = (x+w)^{-n}$, the summation process

$$\sum_{r \in R} \left(\sum_{W'} {}_e f(r+w')\right) = \sum_{r \in R} \left(\sum_v {}_e \sum_\mu {}_e f(r + \mu u' + v v')\right).$$

All these summation processes give the same result when applied to an absolutely convergent series; an essential step in Eisenstein's theory consists in describing how they differ when applied to the series E_1, E_2.

For a moment, let us write $\sum_{e'}$ for any one of these summation processes, and E'_n for the function obtained by applying it to the series $\sum (x+w)^{-n}$. We have $E'_n = E_n$ for $n \geq 3$, since then the series is absolutely convergent. Moreover, since all these processes are compatible with differentiation term by term, we have $dE'_1/dx = -E'_2$, $dE'_2/dx = -2E_3$, so that there are constants A, B (constant, that is to say, with respect to x, but depending upon u and v) such that

(4) $$E'_1 - E_1 = Ax + B, \quad E'_2 - E_2 = -A.$$

Eisenstein, however, prefers to derive this result differently (it is hard to say whether this indicates any lack of confidence in the legitimacy of termwise differentiation); it is enough to indicate his way of dealing with $E_1' - E_1$, since the case of $E_2' - E_2$ is quite similar. One observes that, in calculating $E_1' - E_1$, one may replace a finite number of terms of the series $\sum (x+w)^{-1}$ by 0; this will be done for all the terms for which $|w| \leq M$; denote it by writing $\sum''$ instead of $\sum$; we assume that $|x| < M$. We also write $\sum'$, as usual, for a series where the term corresponding to $\mu = v = 0$ is omitted. We can now write

$$E_1' - E_1 = (\textstyle\sum_{e'}'' - \sum_e'')(x+w)^{-1} = (\textstyle\sum_{e'}'' - \sum_e'')\left(\frac{1}{w} - \frac{x}{w^2} + \frac{x^2}{w^3} - \cdots\right).$$

But it is easily seen that the series $\sum |x^n w^{-n-1}|$, extended to all $n \geq 2$ and to all w for which $|w| > M$, is absolutely convergent (since M is $> |x|$). Therefore the corresponding terms in the above formula may be disregarded. Restoring now in the summation the finitely many terms for which $0 < |w| \leq M$, we get

$$E_1' - E_1 = \textstyle\sum_{e'}' \frac{1}{w} - \sum_e' \frac{1}{w} - x\left(\sum_{e'}' \frac{1}{w^2} - \sum_e' \frac{1}{w^2}\right).$$

Of course, according to the definition of $\sum_e$, $\sum_e' w^{-1}$ is 0, but this may not hold for $\sum_{e'}'$.

§ 4. Now we take up the case where $\sum_{e'}$ is the process derived from $\sum_e$ by a translation on W; as we have observed, this amounts to calculating $E_n(x + w_0)$ for $w_0 \in W$. As all the ε_n are periodic with the period 1, (2) shows that u is a period for all the E_n. On the other hand, in view of (1), we have, e.g. for any integer $m > 0$:

$$E_1(x + mv) - E_1(x) = \lim u^{-1}\left(\sum_{N+1}^{N+m} \varepsilon_1\left(\frac{x+vv}{u}\right) - \sum_{-N}^{-N-1+m} \varepsilon_1\left(\frac{x+vv}{u}\right)\right).$$

We calculate the right-hand side by applying (14) of Chap. II, § 7, and observing that we have, with the notations of § 1:

$$\mathbf{e}\left(\frac{x+vv}{u}\right) = \mathbf{e}\left(\frac{x}{u} + v\delta\tau\right) = \mathbf{e}\left(\frac{x}{u}\right)q^{v\delta}.$$

As $|q| < 1$, this tends to 0 for $v\delta \to +\infty$ and to ∞ for $v\delta \to -\infty$. Therefore, by (14) of Chap. II, $\varepsilon_1\left(\frac{x+vv}{u}\right)$ tends to $\pm \pi i$ according as $v\delta$ tends to $-\infty$ or to $+\infty$. The above formula gives now

$$E_1(x+mv) - E_1(x) = -\frac{2\pi i \delta m}{u}.$$

Of course this implies that the same formula holds for $m \leq 0$. Finally, we get:

$$(5) \qquad E_1(x+\mu u+vv) = E_1(x) - \frac{2\pi i \delta v}{u}.$$

Differentiating this, we see that E_2 (as well as all the E_n for $n \geq 3$) is periodic with the period lattice W.

§ 5. Next, we consider the other summation processes mentioned in § 3; it will be enough to discuss the last one, which contains the others as special cases. Let therefore u', v' be generators for a sublattice W' of W. In matrix notation, we can write

$$(6) \qquad (u' \ v') = (u \ v) \cdot \begin{pmatrix} a & b \\ c & d \end{pmatrix},$$

where a,b,c,d are integers with a determinant $N = ad - bc \neq 0$. Notations being as in § 1, we put $\delta' = \delta(u',v')$; then we have $\delta' = \delta \operatorname{sgn} N$. The lattice W' has the index $|N|$ in W, so that a set R of representatives of W/W' in W consists of $|N|$ elements; it is convenient to assume, once and for all, that 0 is one of them; then, if $W' = W$, we have $R = \{0\}$.

Applying to E_1 the summation process described in § 3, we get the function E_1' given by

$$E_1' = \sum_{r \in R} E_1(x+r; u',v').$$

In view of (4), $E_1' - E_1$ can be written as $Ax + B$, and we must now calculate A and B. To do this, we substitute, first $x+u$, then $-x$ for x; as E_1 has the period u and is an odd function, we get

$$Au = \sum_r E_1(x+u+r; u',v') - \sum_r E_1(x+r; u',v'),$$

$$2B = -\sum_r E_1(x-r; u',v') + \sum_r E_1(x+r; u',v').$$

For each r, we can write $-r = r' + w_r'$, $r+u = r'' + w_r''$, with r', r'' in R and w_r', w_r'' in W'; the r' are a permutation of R, and so are the r''. Put now

$$w_r' = \mu_r u' + v_r v', \qquad w_r'' = \mu_r' u' + v_r' v';$$

in view of (5), the above formulas give:

$$Au = -\frac{2\pi i \delta'}{u'} \sum v_r', \qquad 2B = \frac{2\pi i \delta'}{u'} \sum v_r.$$

At the same time, we have

$$0 = \sum(r+r'+w_r) = 2\sum r + (\sum \mu_r)u' + (\sum v_r)v',$$
$$0 = \sum(r''+w_r''-r-u) = \sum(w_r''-u) = (\sum \mu_r')u' + (\sum v_r')v' - |N|u.$$

In this last formula, we substitute for u its value taken from (6); as δ' is $\delta \operatorname{sgn} N$, this gives

$$\sum v_r' = -\frac{c\delta}{\delta'}.$$

Finally we get:

(7) $$\sum_{r \in R} E_1(x+r; u', v') = E_1(x; u, v) + \frac{2\pi i \delta c x}{uu'} - \frac{\pi i \delta' \bar{v}}{u'}$$

with $\bar{v}$ given by

$$2\sum r = \bar{\mu} u' + \bar{v} v'.$$

Naturally this applies in particular to the case where $|N|=1$, $W'=W$, $R=\{0\}$.

§6. In §2 we postponed the proof of the convergence of (2); this will be given now. Put $\zeta = x/u$, $z = \mathbf{e}(\zeta)$; take first the case $n=1$, and express ε_1 by means of (14) of Chap. II, §7. As we have $v/u = \pm \tau$ and $q = \mathbf{e}(\tau)$, the general term of the series in (2) can be written as

$$\varepsilon_1\left(\frac{x+vv}{u}\right) + \varepsilon_1\left(\frac{x-vv}{u}\right) = \pi i \left(\frac{q^v z + 1}{q^v z - 1} + \frac{q^{-v} z + 1}{q^{-v} z - 1}\right)$$
$$= -2\pi i \left(\frac{1}{1-q^v z} - \frac{1}{1-q^v z^{-1}}\right);$$

for large v, this term, in absolute value, is $< C|q|^v$, with C depending upon z but not upon v; therefore the series (2) is absolutely convergent. For $n \geq 2$, we rewrite (2) as

$$E_n(x) = u^{-n} \sum_v \varepsilon_n\left(\frac{x+vv}{u}\right) = u^{-n} \sum_v \varepsilon_n(\zeta + v\tau),$$

and we will show that here the summation process $\sum_e$ is unnecessary, the series being absolutely convergent. In fact, we have

$$\varepsilon_n(\zeta) = \frac{(-1)^{n-1}}{(n-1)!} \frac{d^{n-1}}{d\zeta^{n-1}} \varepsilon_1(\zeta) = \frac{(-2\pi i)^n}{(n-1)!} \left(z\frac{d}{dz}\right)^{n-1}\left(\frac{1}{1-z}\right).$$

If here we substitute $\zeta + v\tau$ for ζ, this amounts to substituting $q^v z$ for z if $v > 0$ or $(q^{|v|} z^{-1})^{-1}$ for z if $v < 0$. Take first $v > 0$; for v large, we have $|q^v z| < 1$, and we may expand in a power-series in $q^v z$. Thus:

$$\varepsilon_n(\zeta + v\tau) = \frac{(-2\pi i)^n}{(n-1)!} \left(z \frac{d}{dz}\right)^{n-1} \left(\frac{1}{1-q^v z}\right) = \frac{(-2\pi i)^n}{(n-1)!} \sum_{d=1}^{\infty} d^{n-1} q^{vd} z^d.$$

For v large, this is again $< C|q|^v$ in absolute value, with C depending upon z but not upon v. We get a similar formula for $v < 0$; in fact, replacing v by $-v$, we get, when v is so large that $|q^v z^{-1}| < 1$:

$$\varepsilon_n(\zeta - v\tau) = \frac{(2\pi i)^n}{(n-1)!} \sum_{d=1}^{\infty} d^{n-1} q^{vd} z^{-d};$$

this completes the convergence proof.

§ 7. Combining the above formulas, we get, for all $n \geq 1$, the following series for $E_n(x)$:

(8)
$$E_n(x) = u^{-n} \sum_{v=-N}^{+N} \varepsilon_n(\zeta + v\tau)$$
$$+ \frac{(2\pi/iu)^n}{(n-1)!} \sum_{v=N+1}^{+\infty} \sum_{d=1}^{+\infty} d^{n-1} q^{vd} [z^d + (-1)^n z^{-d}],$$

where the double series is easily seen to be absolutely convergent provided N has been taken such that $|q^{N+1} z| < 1$ and $|q^{N+1} z^{-1}| < 1$. In particular, if $|q| < |z| < |q|^{-1}$, we may take $N = 0$.

From this one deduces important formulas for the coefficients of the power-series expansion of E_1 at $x = 0$. As E_1 is an odd function, we can write

$$E_1(x) = \frac{1}{2} [E_1(x) - E_1(-x)] = \frac{1}{x} + \frac{1}{2} \sum_e' \left(\frac{1}{x+w} - \frac{1}{-x+w}\right).$$

Take $|x| < |w|$ for all $w \in W$ except $w = 0$; then this gives

$$E_1(x) = \frac{1}{x} - \sum_e' \left(\sum_{m=1}^{\infty} \frac{x^{2m-1}}{w^{2m}}\right).$$

Here the partial sum

$$\sum_w' \sum_{m=2}^{\infty} \frac{x^{2m-1}}{w^{2m}} = \sum_{m=2}^{\infty} \left(\sum_w' w^{-2m}\right) x^{2m-1}$$

is absolutely convergent; subtracting it from $E_1(x)$, we see that $\sum'_e w^{-2}$ is meaningful, and we get:

(9) $$E_1(x) = \frac{1}{x} - \sum_{m=1}^{\infty} e_m x^{m-1},$$

where the coefficients e_m are 0 for odd m, and otherwise are given by

$$e_2 = \sum'_e w^{-2}; \quad e_{2m} = \sum' w^{-2m} \quad (m \geq 2).$$

Differentiating, we get the power-series for E_n:

(10) $$E_n(x) = \frac{1}{x^n} + (-1)^n \sum_{m=1}^{\infty} \binom{2m-1}{n-1} e_{2m} x^{2m-n};$$

for all $m \geq 1$, e_{2m} is the value of $E_{2m}(x) - x^{-2m}$ for $x=0$. In (8), take $n=2m$; for small x, z is close to 1, so that in (8) we may take $N=0$. For $\zeta=0$, $\varepsilon_{2m}(\zeta) - \zeta^{-2m}$ has the value γ_{2m} given by (16), Chap. II, § 7. Therefore, putting now $x=0$, $\zeta=0$, $z=1$ in (8), we get:

(11) $$e_{2m} = \frac{(2\pi i/u)^{2m}}{(2m-1)!} \left(\frac{(-1)^m B_m}{2m} + 2 \sum_{N=1}^{\infty} \sigma_{2m-1}(N) q^N \right)$$

where $\sigma_{2m-1}(N)$ is defined as usual as the sum of the $(2m-1)$-th powers of the divisors of N:

$$\sigma_{2m-1}(N) = \sum_{d|N} d^{2m-1}.$$

For e_{2m}, Eisenstein writes $(2m, *)$; he writes $\sum^*$ instead of our $\sum'$. When necessary, we shall write $e_{2m}(u,v)$ instead of e_{2m}; for $m \geq 2$, $e_{2m}(u,v)$ depends only upon the lattice W and may be written as $e_{2m}(W)$.

§ 8. Differentiate $n-1$ times, with respect to x, formula (7) of § 5. For $n=2$, we get

(12) $$\sum_{r \in R} E_2(x+r; u', v') = E_2(x; u, v) - \frac{2\pi i \delta c}{u u'},$$

and, for $n \geq 3$:

(13) $$\sum_{r \in R} E_n(x+r; u', v') = E_n(x; u, v)$$

(a trivial formula, since in that case the series are absolutely convergent). For $n=2m$, subtract x^{-n} from both sides and put $x=0$; we get

(14) $$e_2(u',v') + \sum_{r \in R'} E_2(r; u',v') = e_2(u,v) - \frac{2\pi i \delta c}{uu'},$$

(15) $$e_{2m}(u',v') + \sum_{r \in R'} E_{2m}(r; u',v') = e_{2m}(u,v) \quad (m \geq 2)$$

where we have put $R' = R - \{0\}$; we remind that 0 was assumed to belong to the set R of representatives of W/W'. For $W' = W$, (14) gives

(16) $$e_2(u',v') = e_2(u,v) - \frac{2\pi i \delta c}{uu'}.$$

Finally, we observe that we may write

$$E_2(x) - e_2 = x^{-2} + {\sum_e}'[(x+w)^{-2} - w^{-2}],$$

and that here the series is absolutely convergent, so that ${\sum_e}'$ may be replaced by $\sum'$. This is Weierstrass' $\wp$-function, first introduced by him in his lectures of 1862. Also his function ζ is nothing else than $E_1(x) - e_2 x$.

Chapter IV

Basic Relations and Infinite Products

§ 1. In order to obtain the basic relations between the functions $E_n(x)$, Eisenstein uses the same method which he has applied to trigonometric functions (cf. Chap. II). One cannot, however, apply directly the identity (3) of Chap. II, § 2, to the functions E_m defined in (3), Chap. III, § 2; difficulties of convergence would stand in the way. No such difficulty arises if one starts from the identity (2) of Chap. II, § 2, with $m=n=3$; this is the procedure followed at first by Eisenstein. As he indicates later, one may also proceed in two stages, first deriving an identity for trigonometric functions as in Chap. II, and then applying this to the elliptic functions; this is somewhat simpler and will be done now.

As in Chap. III, § 6, we rewrite the definition of E_n as follows:

(1) $$E_n(x) = u^{-n} \sum_{e} \varepsilon_n(\zeta + v\tau)$$

with $\zeta = x/u$, $\tau = \delta v/u$ as before. For convenience, we also rewrite (4) of Chap. II, § 2, with $\zeta, \zeta', \zeta'' = \zeta + \zeta'$ substituted for x, y, z:

(2) $$\varepsilon_2(\zeta)\varepsilon_2(\zeta') - \varepsilon_2(\zeta)\varepsilon_2(\zeta'') - \varepsilon_2(\zeta')\varepsilon_2(\zeta'') = 2\varepsilon_3(\zeta'')[\varepsilon_1(\zeta) + \varepsilon_1(\zeta')].$$

Here we substitute $\zeta + v\tau$ for ζ, $\zeta' + (\rho - v)\tau$ for ζ', and consequently $\zeta'' + \rho\tau$ for ζ''; then we apply the summation process $\sum_{v}^{e}$ to both sides, ρ being kept fixed, and finally we apply $\sum_{\rho}$ to both sides. As we have seen in Chap. III, § 6, the series (1) is absolutely convergent for $n \geq 2$; therefore the summations with respect to v and to ρ can be performed independently on all the terms in the left-hand side of (2) and may be replaced by summations with respect to v and $\rho - v$, or to ρ and $\rho - v$. Consequently, if we write x, x', x'' for $\zeta u, \zeta' u, \zeta'' u$, the result of the double summation on the left-hand side of (2) will be:

$$u^4 [E_2(x) E_2(x') - E_2(x) E_2(x'') - E_2(x') E_2(x'')].$$

On the other hand, the result of the summation $\sum_v$ on the right-hand side is:

$$2u\varepsilon_3(\zeta''+\rho\tau)[E_1(x)+E_1(x'+\delta\rho v)],$$

which, in view of (5), Chap. III, § 5, can be written as

$$2u\varepsilon_3(\zeta''+\rho\tau)\left[E_1(x)+E_1(x')-\frac{2\pi i\rho}{u}\right].$$

Applied to the first two terms, the summation with respect to ρ gives

$$2u^4 E_3(x'')[E_1(x)+E_1(x')].$$

As to the last term, note that we have

$$-2\rho\varepsilon_3(\zeta''+\rho\tau) = \frac{d}{d\tau}\varepsilon_2(\zeta''+\rho\tau).$$

Apply summation with respect to ρ; in view of the formulas and estimates in Chap. III, § 6, everything is absolutely convergent, and we get

$$-2\sum_\rho \rho\varepsilon_3(\zeta''+\rho\tau) = \frac{d}{d\tau}\sum_\rho \varepsilon_2(\zeta''+\rho\tau).$$

Now we regard x, x', u, v as the independent variables, and interpret the symbols for partial differentiation accordingly; our final result is that we have, for $x''=x+x'$:

(3)
$$\begin{aligned}E_2(x)E_2(x')-E_2(x)E_2(x'')-E_2(x')E_2(x'')\\ = 2E_3(x'')[E_1(x)+E_1(x')]+\frac{2\pi i\delta}{u}\frac{\partial E_2(x'')}{\partial v}.\end{aligned}$$

It is analogous to the corresponding formulas for trigonometric functions and for rational functions, except for the occurrence of the last term.

§ 2. We now proceed as in Chap. II, § 3. For a given x, not in W, both sides of (3), regarded as functions of x', have a double pole at $x'=0$; expanding them, one finds that the terms in x'^{-2} and in x'^{-1} are the same; equating the constant terms, we get:

(4) $$\frac{2\pi i\delta}{u}\frac{\partial E_2}{\partial v} = 3E_4 - 2E_1 E_3 - E_2^2.$$

Similarly, for a given x, consider both sides of (3) as functions of x'' and expand at $x''=0$; equating the constant terms gives

(5) $$E_4 = E_2^2 - 2e_2 E_2 - \frac{2\pi i \delta}{u} \frac{\partial e_2}{\partial v}.$$

Put $e_2' = \partial e_2/\partial v$, and expand both sides of (5) at $x=0$; we get

(6) $$\frac{2\pi i \delta}{u} e_2' = 5e_4 - e_2^2,$$

so that we can rewrite (5) as follows:

(7) $$E_4 = (E_2 - e_2)^2 - 5e_4.$$

Since $d^2 E_2/dx^2 = 6E_4$, we recognize here the familiar equation $\wp'' = 6\wp^2 - \frac{1}{2}g_2$ for the "Weierstrass function" $\wp = E_2 - e_2$, with $g_2 = 60e_4$. By integration (or, as Eisenstein chooses to do, by successive differentiation and elimination) one gets from this the first order equation for $\wp$ in Weierstrass' canonical form; Eisenstein writes this (fifteen years before Weierstrass' first lectures on the subject) as follows:

(8) $$E_3^2 = (E_2 - e_2)^3 - 15e_4(E_2 - e_2) + 10(c - e_2 e_4),$$

with (as he says) the "peculiar constant"

$$c = -\frac{\pi i \delta}{2u} \frac{\partial e_4}{\partial v}.$$

One can also find the constant coefficient in the right-hand side of (8) by expanding both sides at $x=0$; one thus finds for it the value $-35e_6$; if one writes $\wp'^2 = 4\wp^3 - g_2 \wp - g_3$ in Weierstrass' notation, this gives $g_3 = 140 e_6$.

Another important relation is best obtained by integrating both sides of (4), regarded as functions of x; up to an additive constant, this gives the formula

(9) $$\frac{2\pi i \delta}{u} \frac{\partial E_1}{\partial v} = E_3 - E_1 E_2;$$

as both sides are odd functions of x, the additive constant must be 0, so that (9) is valid as written above.

As in Chap. II, §4, we will also derive from (3) an addition formula for E_1, which is as follows:

(10) $$(E_2(x) - E_2(x')) \cdot (E_1(x+x') - E_1(x) - E_1(x')) + E_3(x) - E_3(x') = 0.$$

To prove this, write $F(x,x')$ for the left-hand side. Now substitute $-x', x'', x$ for x, x', x'', respectively, in (3), and then use (4) to eliminate $\partial E_2/\partial v$; one gets

$$E_2(x')E_2(x'') - E_2(x')E_2(x'') - E_2(x'')E_2(x) + 2E_3(x)(E_1(x') - E_1(x''))$$
$$-3E_4(x) + 2E_1(x)E_3(x) + E_2(x)^2 = 0.$$

It is easily verified that the left-hand side is nothing else than $\partial F/\partial x$, which is therefore 0. As $F(x,x')$ changes sign when one exchanges x and x', it must be 0.

§ 3. Now, in full analogy with his theory of trigonometric functions, Eisenstein introduces the infinite products

$$f(t,x) = \prod_e \left(1 - \frac{t}{x+w}\right), \quad \varphi(x) = x\prod_e' \left(1 - \frac{x}{w}\right).$$

In order to justify this, e.g. for φ, he observes that, after discarding the factor x and the finitely many factors for which $|w| \leq |x|$, $\log\varphi$ can be written as

$$-\sum_e \left(\sum_{n=1}^{+\infty} \frac{1}{n} (x/w)^n\right);$$

here the double series consisting of the terms for which $|w| > |x|$ and $n \geq 3$ is absolutely convergent, while $\sum_e$ has previously been found to be meaningful for the terms corresponding to $n=1$ and to $n=2$. Incidentally, this amounts to saying that the "Weierstrass canonical product"

$$x\prod' \left(1 - \frac{x}{w}\right) \exp\left(\frac{x}{w} + \frac{x^2}{2w^2}\right),$$

which, in the Weierstrass theory, defines the σ-function, is absolutely convergent, and then using the earlier results on $\sum_e' w^{-1}, \sum_e' w^{-2}$; as these are respectively equal to 0 and to e_2, this gives $\varphi = \sigma \cdot \exp(-e_2 x^2/2)$.

When necessary, we shall write $f(t,x;u,v)$, $\varphi(x;u,v)$ instead of $f(t,x)$, $\varphi(x)$. The relation between these functions and the function E_1 is obvious; it is given by

(11) $$E_1(x) = \frac{d}{dx}\log\varphi(x), \quad f(t,x) = \frac{\varphi(x-t)}{\varphi(x)}, \quad \varphi(t) = -[xf(t,x)]_{x=0}.$$

In view of the definition of $\prod_e$, we can write

(12) $$f(t,x) = \prod_{e \atop v} \prod_{e \atop \mu} \left(1 - \frac{t}{x + \mu u + v v}\right).$$

In view of (13), Chap. II, § 6, the second product has the value

$$P_v = \frac{\sin \pi \dfrac{x-t+vv}{u}}{\sin \pi \dfrac{x+vv}{u}}.$$

As before, we put $\tau = \delta v/u$, $\zeta = x/u$, $q = \mathbf{e}(\tau)$, $z = \mathbf{e}(\zeta)$; also, put $\zeta^* = (x-t)/u$, $z^* = \mathbf{e}(\zeta^*)$, and, for all $v \geq 1$:

$$A_v = 1 - q^v z, \quad B_v = 1 - q^v z^{-1}, \quad A_v^* = 1 - q^v z^*, \quad B_v^* = 1 - q^v z^{*-1}.$$

A trivial calculation gives now

$$P_0 = (z^{*\tfrac{1}{2}} - z^{*-\tfrac{1}{2}})/(z^{\tfrac{1}{2}} - z^{-\tfrac{1}{2}}), \quad P_v P_{-v} = A_v^* B_v^* / A_v B_v \quad (v \geq 1).$$

This suggests introducing the absolutely convergent product:

(13) $$X_q(z) = (z^{\tfrac{1}{2}} - z^{-\tfrac{1}{2}}) \prod_{v=1}^{\infty} (1 - q^v z)(1 - q^v z^{-1}),$$

where, as we said before, we take $z^{\tfrac{1}{2}} = \mathbf{e}(\zeta/2)$. Then we get:

$$f(t,x) = X_q(z^*)/X_q(z).$$

For $x = 0$, $z = 1$, the infinite product in (13) takes the value $P(q)^2$, if we put

(14) $$P(q) = \prod_{v=1}^{\infty} (1 - q^v);$$

at the same time, the first factor in (13) has the expansion

$$\mathbf{e}(x/2u) - \mathbf{e}(-x/2u) = 2\pi i x/u + \cdots.$$

Now (11) gives at once the formula

(15) $$\varphi(x) = \frac{u}{2\pi i} \frac{X_q(z)}{P(q)^2}.$$

§ 4. The convergence proof given in § 3 for the infinite products involved in the definition of $f(t,x)$ implies that the function

$$\log f(t,x) + t \sum_e (x+w)^{-1} + \frac{t^2}{2} \sum_e (x+w)^{-2}$$

remains unchanged when $\sum_e$ is replaced by any one of the alternative summation processes discussed in Chap. III, § 3. Therefore the formulas in Chap. III, §§ 4—5, imply corresponding formulas for $f(t,x)$.

Take in particular the formula (7) in Chap. III, § 5, and the one derived from it by differentiation. This gives

(16) $$\prod_{r \in R} f(t, x+r; u', v') = f(t, x; u, v) \mathbf{e}\left(\frac{\delta c}{2uu'}(t^2 - 2xt) + \frac{\delta' \overline{v} t}{2u'}\right).$$

Multiply this with x, and put $x = 0$; by (11), we get

(17) $$\varphi(t; u', v') \prod_{r \in R'} f(t, r; u', v') = \varphi(t; u, v) \mathbf{e}\left(\frac{\delta c t^2 + \delta' \overline{v} t u}{2uu'}\right),$$

where, as before, we have put $R' = R - \{0\}$.

Assume for instance that $ad - bc = 1$, i.e. that (a, b, c, d) belongs to the modular group; then $W' = W$ and $R' = \emptyset$. To simplify notations, assume that $\delta = 1$, so that $\tau = v/u$, and put:

$$\tau' = \frac{v'}{u'} = \frac{d\tau + b}{c\tau + a}, \quad q' = \mathbf{e}(\tau'), \quad \zeta' = \frac{x}{u'} = \frac{\zeta}{c\tau + a}, \quad z' = \mathbf{e}(\zeta').$$

Writing now x instead of t in (17), and expressing φ by means of (15), we get

(18) $$X_{q'}(z') = X_q(z) \cdot (c\tau + a)^{-1} \frac{P(q')^2}{P(q)^2} \mathbf{e}\left(\frac{c\zeta\zeta'}{2}\right).$$

§ 5. We may also apply formula (5) of Chap. III, § 4, in order to get a corresponding formula for $f(t, x)$; but one gets a better result by using the infinite product for X_q. In fact, changing x into $x + u$ merely changes the sign of $z^{1/2}$ and consequently that of $X_q(z)$. On the other hand, changing x into $x + v$ amounts to replacing z by $q^\delta z$. We can now easily calculate $X_q(q^v z)$ for every v by observing that the absolutely convergent products for $X_q(z)$ and for $X_q(q^v z)$ differ only by a finite number of factors; in view of this, a trivial calculation gives

(19) $$X_q(q^v z) = q^{-v^2/2}(-z)^{-v} X_q(z).$$

Combining this with (15), we get

(20) $$\varphi(x + \mu u + v v) = (-1)^{\mu + v} \varphi(x) \mathbf{e}\left(-\delta v \frac{x}{u} - \delta v^2 \frac{v}{2u}\right).$$

In view of (11), this provides us with an alternative proof for (5) of Chap. III, or at any rate with a variant of the earlier proof.

§ 6. "For lack of space", says Eisenstein, he refrains from including in his presentation of the theory the relations between "theta-series" and infinite products whose discovery had been perhaps, according to Dirichlet, Jacobi's most far-reaching achievement. We shall now repair this omission.

Any absolutely convergent product of the form $\prod(1+a_n)$ can be expanded into an absolutely convergent series by multiplying out its factors. Applying this elementary observation (which seems to go back to Euler) to the product $X_q(z)$, we get an expansion

$$X_q(z) = z^{\frac{1}{2}} \sum_{n=-\infty}^{+\infty} \sum_{\nu=0}^{+\infty} C_{n,\nu} q^\nu z^n = \sum_{n=-\infty}^{+\infty} F_n(q) z^{n+\frac{1}{2}},$$

absolutely convergent for $|q|<1$, $z \neq 0$, where the $C_{n,\nu}$ are rational integers and the $F_n(q)$ are power-series in q, absolutely convergent for $|q|<1$. Writing that this satisfies (19), we get

$$F_{n+\nu}(q) = (-1)^\nu F_n(q) q^{(\nu^2 + \nu + 2n\nu)/2}.$$

Writing F instead of F_0, we get

(21) $\qquad X_q(z) = F(q) T(q, z), \quad T(q, z) = z^{\frac{1}{2}} \sum_{n=-\infty}^{+\infty} (-1)^n q^{(n^2+n)/2} z^n,$

where we still have to determine $F(q)$. From (13) it follows that $F(0)=1$.

In the *Fundamenta*, Jacobi had shown that $F(q)$ is $P(q)^{-1}$. This can be verified for instance, according to Kronecker, by verifying the formula

$$T(q^4, q^2) = T(q, iq^{\frac{1}{2}}) \mathbf{e}\left(\frac{1}{8} - \frac{3\tau}{4}\right)$$

and then expressing T on both sides in terms of X_q by means of (21). One finds then that the function $F(q)P(q)$ is unchanged if one substitutes q^4 for q; as it is a power-series in q, beginning with the constant term 1 and convergent for $|q|<1$, it must be 1.

§ 7. It seems more interesting, however, to derive the same result from the partial differential equations of parabolic type for E_1 and for T. Actually (9) of § 2 is such an equation, since $\partial E_1/\partial x = -E_2$, $\partial E_2/\partial x = -2E_3$; and the fact that a "theta-series" such as $T(q,z)$ satisfies a parabolic equation is well-known; even before Jacobi, theta-series had been introduced into the mathematical literature by Fourier in his work on the heat equation.

As before, we regard x, u, v as the independent variables, or rather (for the time being) we regard u as a constant parameter and x, v as the independent variables, and we write the symbols for partial differentiation accordingly. From (11), § 3, we get

$$E_2 - E_1^2 = -\frac{1}{\varphi} \frac{\partial^2 \varphi}{\partial x^2}, \quad E_3 - E_1 E_2 = \frac{1}{2} \frac{\partial}{\partial x}\left(\frac{1}{\varphi} \frac{\partial^2 \varphi}{\partial x^2}\right).$$

Now (9) of §2 can be written as

$$\frac{\partial}{\partial x}\left(\frac{1}{\varphi}\frac{\partial^2\varphi}{\partial x^2} - \frac{4\pi i\delta}{u\varphi}\frac{\partial\varphi}{\partial v}\right) = 0;$$

here the quantity within brackets must therefore be equal to its value for $x=0$. As the first term is $E_1^2 - E_2$, and as E_1, E_2 have the expansions $x^{-1} - e_2 x + \cdots$, $x^{-2} + e_2 + \cdots$ at $x=0$, its value at $x=0$ is $-3e_2$. As to the other term, it is 0 for $x=0$, since $x^{-1}\varphi(x)$ has the value 1 for $x=0$. This gives

$$\frac{1}{\varphi}\left(\frac{\partial^2\varphi}{\partial x^2} - \frac{4\pi i\delta}{u}\frac{\partial\varphi}{\partial v}\right) = -3e_2.$$

Here we express φ in terms of $T(q,z)$ by means of (15) and (21), and observe that $T(q,z)$ is a solution of the equation

$$\frac{\partial^2 T}{\partial x^2} - \frac{4\pi i\delta}{u}\frac{\partial T}{\partial v} + \frac{\pi^2}{u^2}T = 0.$$

This gives

(22) $$\frac{4\pi i\delta}{u}\frac{\partial}{\partial v}\log(FP^{-2}) = 3e_2 - \frac{\pi^2}{u^2}.$$

Now the definition of $P(q)$ gives at once:

$$\frac{\partial}{\partial v}\log P(q) = \frac{2\pi i\delta}{u}q\frac{d}{dq}\log P(q) = -\frac{2\pi i\delta}{u}\sum_{v=1}^{+\infty}\frac{vq^v}{1-q^v}.$$

A comparison of the right-hand side with (11) of Chap. III, §7 (for $m=1$) shows that this series is the same which occurs in that formula for e_2; as we had found that $\gamma_2 = \pi^2/3$, i.e. $B_1 = \frac{1}{6}$, this gives:

(23) $$e_2 = \frac{\pi^2}{3u^2} + \frac{8\pi^2}{u^2}q\frac{d}{dq}\log P(q) = \frac{\pi^2}{3u^2} - \frac{4\pi i\delta}{u}\frac{\partial}{\partial v}\log P(q)$$
$$= -\frac{4\pi i\delta}{u}\frac{\partial}{\partial v}\log(q^{1/24}P(q)).$$

Substituting in (22), for e_2, the second one of these values, we see that FP^{-2} differs from P^{-3} only by a constant factor; as F and P are both 1 at $q=0$, this factor is 1. We have thus proved again that $F = P^{-1}$.

§ 8. Traditionally, the function $q^{1/24}P(q)$ which occurs in (23) is denoted by $\eta(\tau)$, with $q = \mathbf{e}(\tau)$ as always. Thus we have:

(24) $$e_2 = -\frac{4\pi i}{u^2}\frac{d}{d\tau}\log\eta(\tau), \quad \eta(\tau) = P(\mathbf{e}(\tau))\,\mathbf{e}\!\left(\frac{\tau}{24}\right).$$

The transformation formula (14) of Chap. III, § 8, gives now a transformation formula for η. As in § 4, take (a,b,c,d) in the modular group; assume $\delta=1$, and take the same notations as there. We have $d\tau'/d\tau=(u/u')^2$; taking this into account, one finds at once that (14) of Chap. III can be rewritten as follows:

$$\frac{d}{d\tau}\log\frac{\eta(\tau')}{\eta(\tau)} = \frac{1}{2}\frac{c}{c\tau+a}.$$

The right-hand side is the logarithmic derivative of $(c\tau+a)^{\frac{1}{2}}$, so that $\eta(\tau')/\eta(\tau)$ can differ from this only by a constant factor; if we call this factor ε_A, with $A=(a,b,c,d)$, we may thus write:

(25) $$\eta\left(\frac{d\tau+b}{c\tau+a}\right) = \varepsilon_A \eta(\tau)\cdot(c\tau+a)^{\frac{1}{2}},$$

where, for definiteness, we may assume that $(c\tau+a)^{\frac{1}{2}}$ is the branch of the square root with positive real part. Somewhat more precise results will be given later on; the full determination of the factor ε_A was obtained by Dedekind; this is apparently why η is frequently called "the Dedekind function". For the case $A=(0,-1,1,0)$, ε_A can be determined by substituting i for τ in (25); as η is never 0, this gives $\varepsilon_A = i^{-\frac{1}{2}}$, so that we get:

$$\eta(-1/\tau) = \eta(\tau)\cdot(\tau/i)^{\frac{1}{2}}.$$

Combining now (25) with (18), we get the transformation formula for theta-functions. Here, as had already been noticed by Jacobi, it is convenient to introduce the modified function $i^{-1}q^{\frac{1}{8}}T(q,z)$, regarding this as a function of ζ and τ. With Kronecker[5], we put

(26) $$\theta(\zeta,\tau) = i^{-1}q^{\frac{1}{8}}T(q,z) = \sum_{n=-\infty}^{+\infty} \mathbf{e}\left(\left(n+\frac{1}{2}\right)^2\frac{\tau}{2} + \left(n+\frac{1}{2}\right)\left(\zeta-\frac{1}{2}\right)\right).$$

This satisfies a simpler partial differential equation than T,

$$\frac{\partial^2\theta}{\partial\zeta^2} = 4\pi i \frac{\partial\theta}{\partial\tau},$$

---

[5] For typographical reasons, we write θ for Kronecker's ϑ; Jacobi's notation is slightly different. Our notation θ coincides with Jordan's in Chap. VII of his *Cours d'Analyse*, vol. II; it does not seem to be generally known that this chapter is perhaps the best existing treatment of the theory of elliptic and modular functions along classical lines.

and it has a simpler transformation formula, which follows at once from (18), (21) and (25):

(27) $$\theta(\zeta',\tau') = \varepsilon_A^3\, \theta(\zeta,\tau)\, \mathbf{e}\!\left(\frac{c\zeta\zeta'}{2}\right)\cdot (c\tau+a)^{\frac12}\,;$$

here, as before, $A=(a,b,c,d)$ belongs to the modular group, and we have put

$$\tau' = \frac{d\tau+b}{c\tau+a},\qquad \zeta' = \frac{\zeta}{c\tau+a}.$$

In the special case $\tau' = -1/\tau$, we have found $\varepsilon_A = i^{-\frac12}$; this gives for θ a formula which is usually proved by applying "Poisson summation" to the theta-series; it is worth noting that the proof obtained here by following in Eisenstein's footsteps seems essentially different, and gives a more general result.

§ 9. After reaching this point, it is not hard to obtain most (and perhaps all) of the classical formulas in that theory. Without aiming at completeness, we shall consider some examples.

The simplest ones are obtained by taking what may be regarded as the essential result of §§ 5—6, viz., the formula

$$T(q,z) = P(q)\, X_q(z),$$

and either differentiating this with respect to z at $z=1$, or substituting for z the special values -1, $\pm q^{-\frac12}$ or $q^{\frac13}$. As we have observed, $dX_q(z)/dz$ has the value $P(q)^2$ at $x=0$, $z=1$; accordingly, we have

(27) $$\frac{1}{2}\sum_{n=-\infty}^{+\infty}(-1)^n(2n+1)q^{(n^2+n)/2} = P(q)^3\,;$$

we will write T_0 for the common value of both sides. For $z=-1$, we get

(28) $$T_1 = \frac{1}{2}\sum_{n=-\infty}^{+\infty} q^{(n^2+n)/2} = P(q)\prod_{1}^{\infty}(1+q^\nu)^2 = P(q^2)^2\, P(q)^{-1}\,;$$

similarly, for $z=\pm q^{-\frac12}$ and $z=q^{\frac13}$, we get

(29) $$T_2 = \sum_{n=-\infty}^{+\infty}(-1)^n q^{n^2/2} = P(q)\prod_{0}^{\infty}(1-q^{\nu+\frac12})^2 = P(q^{\frac12})^2\, P(q)^{-1},$$

(30) $$T_3 = \sum_{n=-\infty}^{+\infty} q^{n^2/2} = P(q)\prod_{0}^{\infty}(1+q^{\nu+\frac12})^2 = P(q)^5\, P(q^2)^{-2}\, P(q^{\frac12})^{-2},$$

(31) $$\sum_{n=-\infty}^{+\infty} (-1)^{n+1} q^{(n+1)(3n+2)/6} = P(q^{\frac{1}{3}}).$$

This last formula (with q replaced by x^3) had been discovered by Euler in 1740 and proved by him in 1750.

§ 10. Other formulas can be obtained by substituting $x+x'$ for x' in the addition formula (10) of § 2, and expressing E_1 in terms of φ in that formula by means of (11), § 3. This gives a formula which can be written as follows:

$$\frac{\partial}{\partial x} \log \frac{\varphi(2x+x')}{\varphi(x)^2 \varphi(x+x')^2} = \frac{\partial}{\partial x} \log[E_2(x) - E_2(x+x')].$$

This shows that the functions whose logarithmic derivatives appear on both sides differ only by a factor of the form $f(x')$; to determine this factor, we expand at $x=0$ and equate terms in x^{-2}; this gives, respectively, $\varphi(x')^{-1} x^{-2}$ and x^{-2}, so that $f(x')$ is $\varphi(x')^{-1}$, and therefore:

(32) $$\frac{\varphi(2x+x')\varphi(x')}{\varphi(x)^2 \varphi(x+x')^2} = E_2(x) - E_2(x+x').$$

One would get a more symmetrical formula by substituting here $x'-x$ for x'; one can also express φ in terms of X_q, or of θ. Leaving this aside, we will merely consider the case where $2x$ belongs to the period lattice W. If $x \in W$, both sides of (32) have a pole at x; therefore we take $x = w/2$ with $w = \mu u + \nu v \in W$, $w \notin 2W$. Taking (20) of § 5 into account, and then replacing x' by $x - w/2$, one finds that $E_2(w/2) - E_2(x)$ differs from the function

$$\varphi(x)^{-2} \varphi\left(x - \frac{w}{2}\right)^2 \mathbf{e}\left(-\frac{\delta v x}{u}\right)$$

by a factor independent of x. Comparing the expansions of both at $x=0$, we see that this factor is $-\varphi(w/2)^2$. This gives:

(33) $$(E_2(x) - E_2(w/2))^{\frac{1}{2}} = -\frac{\varphi\left(x - \frac{w}{2}\right)}{\varphi(x)\varphi\left(\frac{w}{2}\right)} \mathbf{e}\left(-\frac{\delta v x}{2u}\right),$$

where the left-hand side is the branch whose expansion at $x=0$ begins with x^{-1}.

If in (33) we substitute $w'/2$ for x, with $w' \in W$, $w' \notin 2W$, we find the quantities

$$(E_2(w'/2) - E_2(w/2))^{\frac{1}{2}}$$

expressed in terms of the quantities of the form $\varphi(w/2)$, or (what amounts to the same, in view of (20), §5) in terms of $\varphi\left(\dfrac{u}{2}\right)$, $\varphi\left(\dfrac{v}{2}\right)$, $\varphi\left(\dfrac{u\pm v}{2}\right)$; using (15) of §3, we can then express the latter in terms of $T(q,z)$ for $z=-1$, $z=q^{\frac{1}{2}}$, $z=-q^{\frac{1}{2}}$, respectively; the calculation is easily carried out, and gives the following result. Put

$$a_1 = E_2\left(\frac{u}{2}\right), \quad a_2 = E_2\left(\frac{v}{2}\right), \quad a_3 = E_2\left(\frac{u+v}{2}\right),$$

$$b_1 = a_3 - a_2, \quad b_2 = a_1 - a_3, \quad b_3 = a_1 - a_2.$$

We apply formulas (28), (29), (30) of §9, observing at the same time that by multiplying out these formulas, one gets

$$T_1 T_2 T_3 = P(q)^3 = T_0.$$

Then (33) gives:

(34) $$b_1 = \frac{16\pi^2}{u^2} q^{\frac{1}{2}} T_1^4, \quad b_2 = \frac{\pi^2}{u^2} T_2^4, \quad b_3 = \frac{\pi^2}{u^2} T_3^4.$$

§ 11. Now we recall formula (8) of §2, which can be written, as we have seen there:

(35) $$E_3^2 = (E_2 - e_2)^3 - 15 e_4 (E_2 - e_2) - 35 e_6.$$

The right-hand side is a polynomial in E_2 whose roots are determined by Eisenstein by observing that E_3 is an odd periodic function, so that

$$E_3\left(\frac{w}{2}\right) = E_3\left(-\frac{w}{2}\right) = -E_3\left(\frac{w}{2}\right) = 0$$

whenever $w \in W$, $w \notin 2W$. Consequently a_1, a_2, a_3 are roots of the right-hand side of (35); as the b_i are $\neq 0$, this gives

$$E_3^2 = (E_2 - a_1)(E_2 - a_2)(E_2 - a_3).$$

The discriminant of the right-hand side can thus be written as

$$(b_1 b_2 b_3)^2 = 4(15 e_4)^3 - 27(35 e_6)^2.$$

With an added numerical factor 2^4 which is a matter of tradition and convenience, this defines the classical "discriminant" Δ, which, in view of the formulas given above, is thus:

(36) $$\Delta = g_2^3 - 27 g_3^2 = 2^4 \, 3^3 \, 5^2 (20 e_4^3 - 49 e_6^2) = \left(\frac{2\pi}{u}\right)^{12} q P(q)^{24} = \left(\frac{2\pi}{u}\right)^{12} \eta^{24}.$$

Here e_4, e_6 and therefore Δ depend only upon the period lattice W; therefore, in the transformation formula (25) for η, the factor ε_Δ is a 24-th root of unity. For Δ, we shall also write $\Delta(W)$, or $\Delta(u,v)$, whenever convenient.

Chapter V
Variation I

§ 1. Eisenstein's major themes, properly modulated, lend themselves to a large number of interesting variations; as we indicated earlier (Chap. I; cf. Chap. II, § 5), much of Kronecker's best work consists of such variations, although Kronecker could of course not refrain from adding some themes of his own to Eisenstein's; this will be discussed in Chap. VII and VIII. In this chapter and the next one, we will stay closer to Eisenstein; as an example of the scope of his ideas, we will include a proof for a valuable result, due to R. M. Damerell[6], which in turn has supplied a starting point for some recent investigations of Yu. Manin and S. Vishik[7].

One virtue of Eisenstein's approach is that it supplies directly (without recourse to function-theory) so many of the formulas in the theory of elliptic functions, in the explicit form which is most appropriate for their use in number-theory. We begin by adding a few formulas to those already obtained in Chap. III and IV.

Substituting in (7) of Chap. IV, § 2, the power-series expansions for E_2, E_4, given by (10) of Chap. III, § 7, we get the recursion formula for the coefficients e_{2m}:

$$(1) \qquad \tfrac{1}{3}(m-3)(4m^2-1)e_{2m} = \sum_{r=2}^{m-2}(2r-1)(2m-2r-1)e_{2r}e_{2m-2r} \qquad (m \geq 4),$$

which shows that all the e_{2m} for $m \geq 2$ are in the ring $\mathbf{Q}[e_4, e_6]$.

Similarly, substituting the power-series expansions for E_1, E_2, E_3 into (9), Chap. IV, § 2, we get

$$(2) \qquad \frac{2\pi i \delta}{u} \frac{\partial e_{2m}}{\partial v} = m(2m+3)e_{2m+2} - m\sum_{r=1}^{m} e_{2r}e_{2m-2r+2} \qquad (m \geq 1);$$

---

[6] *Acta Arithmetica* 17 (1970), p. 287.
[7] To appear in *Mat. Sbornik*.

by induction on λ, this shows that we have

$$\left(\frac{2\pi i}{u}\right)^\lambda \frac{\partial^\lambda e_{2m}}{\partial v^\lambda} \in \mathbf{Z}[e_2, e_4, \ldots, e_{2m+2\lambda}]$$

for all $m \geq 1$ and all $\lambda \geq 0$.

§ 2. From the point of view of function-theory, it is clear that every meromorphic function, periodic with the period lattice W and with no poles outside W, must be, up to an additive constant, a linear combination of the functions E_n for $n \geq 2$; that is a simple consequence of Liouville's theorem. This can be applied for instance to any polynomial in $E_2, \ldots, E_n$. It is worth noting that the latter result is also an immediate consequence of Eisenstein's method. Take first the case of $E_m E_n$ with $m \geq 3$, $n \geq 3$; we calculate this by putting $p = x + w$, $r = w'$, $q = -x - w + w'$ in the identity (2), Chap. II, § 2, and performing, first $\sum_e$ with respect to w, and then $\sum'$ with respect to w'. As m, n are ≥ 3, everything is absolutely convergent except the terms corresponding to $h = m - 1$ and $h = m - 2$, and to $k = n - 1$ and $k = n - 2$, in the right-hand side; as to the latter, one has to make use of formula (5) of Chap. III, § 4. One gets in this manner

$$(-1)^n (E_m E_n - E_{m+n}) = \sum_{h=0}^{m-1} \binom{n+h-1}{h} E_{m-h} e_{n+h}$$
$$+ (-1)^n \sum_{k=0}^{n-1} (-1)^k \binom{m+k-1}{k} E_{n-k} e_{m+k} + \frac{C}{u} \frac{\partial e_{m+n-2}}{\partial v}$$

where C is a numerical constant. We have $e_r = 0$ unless r is even; consequently, in the second sum, only those terms for which $(-1)^k = (-1)^m$ are not 0. As was to be expected, the terms in E_1 always cancel each other. From this we can subtract the formula obtained by expanding at $x = 0$ on both sides and equating the constant terms. This gives the more convenient formula

$$(E_m - e_m)(E_n - e_n) - (E_{m+n} - e_{m+n})$$
(3)
$$= (-1)^n \sum_{h=1}^{m-2} \binom{n+h-1}{h} e_{n+h}(E_{m-h} - e_{m-h})$$
$$+ (-1)^m \sum_{k=1}^{n-2} \binom{m+k-1}{k} e_{m+k}(E_{n-k} - e_{n-k}) + (-1)^m \binom{m+n}{m} e_{m+n}.$$

The same formula is valid for $m = 2$, $n \geq 2$; this may be verified by applying the procedure used in Chap. IV, § 1, i.e. by deriving first a formula for the functions ε_n and then the one for E_n. A simpler procedure consists in substituting x, $-x-t$ and $-t$ for x, x', x'' in (3), Chap. IV, § 1, expanding both sides in power-series in t at $t = 0$ and equating the two series. For $m = n = 2$, one gets once more formula (7) of Chap. IV, § 2. Of course the results are the same as what one would get by using function-theory.

One will note that e_2 never occurs in (3) except in the combination $E_2 - e_2$. Putting $m=2$ in (3), one may regard this as an inductive formula expressing E_{n+2} in terms of $E_2 - e_2$ and of the functions E_m for $3 \leq m \leq n$, and ultimately in terms of $E_2 - e_2$ and E_3, with coefficients in $\mathbf{Z}[e_4, e_6, e_8, \ldots]$. On the other hand, it also shows that $1, E_2 - e_2, E_3, E_4, \ldots$, make up a basis for the ring generated by $E_2 - e_2$ and E_3 over $\mathbf{Z}[e_4, e_6, e_8, \ldots]$.

One will also note that the functions $E_2 - e_2$ and E_n for $n \geq 3$, and the quantities e_{2m} for $m \geq 2$, depend only upon the lattice W, and not upon the choice of the generators u, v for W. It will be convenient to write:

$$e(W) = \{e_4, e_6, e_8, \ldots\}; \qquad E(W) = \{1, E_2 - e_2, E_3, E_4, \ldots\}.$$

Then (1) shows that the field of quotients of the ring $\mathbf{Z}[e(W)]$ is $\mathbf{Q}(e_4, e_6)$, and (3) shows that $E(W)$ is a basis for the ring $\mathbf{Z}[e(W), E(W)]$ over $\mathbf{Z}[e(W)]$. For reasons of typographical convenience, we will sometimes write e_W, E_W instead of $e(W), E(W)$; when only one lattice W is being considered, we will also write simply e for $e(W)$, and E for $E(W)$. We draw attention to the fact that, in this notation, e_2 is not in $\mathbf{Q}(e)$ and E_2 not in $\mathbf{Q}(e, E)$.

§ 3. The formulas in Chap. III, § 8, contain as special cases the formulas for the multiplication and division of elliptic functions. For (a,b,c,d), take $(N,0,0,N)$, where N is any integer >1; then $W' = NW$, and we may take for R the set $\{r\}$ consisting of the elements $r = \mu u + \nu v$ with $0 \leq \mu, \nu < N$. From (12) and (13), Chap. III, § 8, we get, for all $n \geq 2$:

$$\sum_{r \in R} E_n(x+r; Nu, Nv) = E_n(x; u, v).$$

As E_n is homogeneous of degree $-n$ in x, u, v, this gives:

(4) $$\sum_{r \in R} E_n\left(\frac{x+r}{N}\right) = N^n E_n(x),$$

where we have simplified the notation by omitting u, v. This is a division formula for E_n; one gets a multiplication formula by substituting Nx for x.

It is important to observe that, for $n=2$, the number of terms in the left-hand side is N^2, so that the formula remains valid if one replaces E_2 by $E_2 - e_2$.

Take now any F in the ring $\mathbf{Z}[e, E]$; as we have shown in § 2, this can be expressed in terms of the basis $E = E(W)$, with coefficients in $\mathbf{Z}[e]$; as every element of that basis satisfies a formula of type (4), this shows that the sum

$$\sum_{r \in R} F\left(\frac{x+r}{N}\right)$$

is in the ring $\mathbf{Z}[e, E]$; for a given F, this sum can be expressed explicitly in terms of the basis E by means of the above formulas.

Applying this result to the powers F^m for $1 \leq m \leq N^2$, we conclude that $F\left(\dfrac{x}{N}\right)$ and all the functions $F\left(\dfrac{x+w}{N}\right)$ with $w \in W$ are algebraic over $\mathbf{Q}(e, E)$ and integral over the ring $\mathbf{Q}[e, E]$. An investigation, by the same methods, of the Galois group of the extension of $\mathbf{Q}(e, E)$ generated by elements of the form $F\left(\dfrac{x}{N}\right)$ would carry us too far; among other things, it would require (as Jacobi had observed in connection with Abel's work) the introduction of Lagrangian resolvents for the Abelian extensions of $\mathbf{C}(e, E)$ generated by elements $F\left(\dfrac{x}{N}\right)$; these are the same functions which have already been alluded to in Chap. II at the end of § 5; they are special cases of the functions to be studied in Chap. VIII.

§ 4. Subtract $(N/x)^n$ from both sides in (4), and put $x = 0$; one gets

(5) $$\sum_{r \in R'} E_n(r/N) = (N^n - 1) e_n,$$

with $e_n = 0$, as we know, for n odd; as before, we have put $R' = R - \{0\}$. For $n = 2$, this can be written as

(6) $$\sum_{r \in R'} (E_2(r/N) - e_2) = 0.$$

Take again any F in $\mathbf{Z}[e, E]$; proceeding as in § 3, we see, firstly, that the sum $\sum F(r/N)$, taken over $r \in R'$, is in $\mathbf{Z}[e]$, and then that $F(w/N)$, for all w in W and not in NW, is algebraic over $\mathbf{Q}(e)$ and integral over $\mathbf{Q}[e]$.

§ 5. Now, as in Chap. III, § 5 and § 8, we consider the general "transformation"

(7) $$(u' \ v') = (u \ v) \begin{pmatrix} a & b \\ c & d \end{pmatrix}$$

of determinant $N = ad - bc$. Put $u'' = Nu$, $v'' = Nv$; write W, W', W'' for the lattices generated respectively by u, v, by u', v', and by u'', v''. Then W' is of index $|N|$ in W, and W'' is of index $|N|$ in W'. Call S a set of representatives for W'/W'' in W', containing 0, and put $S' = S - \{0\}$.

Let us write $e_{2m}, e'_{2m}, e''_{2m}$ for $e_{2m}(u, v)$, $e_{2m}(u', v')$ and $e_{2m}(u'', v'')$, respectively; as e_{2m} is homogeneous of degree $-2m$ in u, v, we have $e''_{2m} = N^{-2m} e_{2m}$. Now formula (15) of Chap. III, § 8, applied to the lattices W', W'', gives (in view of the homogeneity of E_{2m}):

$$N^{2m} e'_{2m} = e_{2m} + \sum_{s \in S'} E_{2m}(s/N) \quad (m \geq 2).$$

In view of our results in §4, this shows that e'_{2m} is integral over $\mathbf{Q}[e]$; similarly, e''_{2m}, and therefore e_{2m}, are integral over $\mathbf{Q}[e']$. More generally, let W_1, W_2 be two "commensurable" lattices; this means that their intersection W_3 is of finite index both in W_1 and in W_2, which is the case if and only if $\mathbf{Q}W_1 = \mathbf{Q}W_2$. Applying what we have just proved, first to W_1 and W_3 and then to W_2 and W_3, we see that $\mathbf{Q}[e(W_1), e(W_2)]$ is integral both over $\mathbf{Q}[e(W_1)]$ and over $\mathbf{Q}[e(W_2)]$. Combining this with the results of §4, we see that, for any F in $E(W_2)$ or in $\mathbf{Z}[e(W_2), E(W_2)]$, the values of F at all points of $\mathbf{Q}W_2$, not in W_2, are integral over the ring $\mathbf{Q}[e(W_1)]$.

Take again W and W' as above, and take any F in $E(W')$. If $F = E_n(x; u', v')$ with $n \geq 3$, formula (13) of Chap. III, §8, gives

$$\sum_{r \in R} F(x+r) = E_n(x; u, v).$$

If $F = E_2(x; u', v') - e_2(u', v')$, take formulas (12) and (14) of Chap. III, §8, and subtract the latter from the former; this gives

$$\sum_{r \in R} F(x+r) - \sum_{r \in R'} F(r) = E_2(x; u, v) - e_2(u, v).$$

Write f_2 for the second sum in the left-hand side; this is integral over $\mathbf{Q}[e(W)]$. Now take any F in $\mathbf{Z}[e(W'), E(W')]$, and express it in terms of the basis $E(W')$; our formulas show that the sum $\sum F(x+r)$, taken over $r \in R$, is a linear combination of f_2 and of the elements of $E(W)$, with coefficients in $\mathbf{Z}[e(W')]$. As this applies to all the powers F^m of F, and as f_2 and $e(W')$ are integral over $\mathbf{Q}[e(W)]$, this proves that $F(x)$ and all the functions $F(x+w)$ with $w \in W$ are integral over the ring $\mathbf{Q}[e(W), E(W)]$.

§6. As Eisenstein observes, a specially noteworthy case occurs when the "transformed" lattice W' can be written as ωW with $\omega \in \mathbf{C}^\times$; one can then choose for W' the generators $u' = \omega u$, $v' = \omega v$, and ω is a characteristic root of the linear substitution (7). Clearly, if ω is real, ωW cannot be a sublattice of W unless ω is a rational integer; if that is not so, ω must be an integer in an imaginary quadratic field, and one says that W admits the complex multiplier ω; this is the phenomenon first discovered by Abel. If we put $N = ad - bc$ as before, we have $N = \omega \bar{\omega} > 0$. If $\omega W = W$, ω must be either a fourth root of unity (this is the so-called "lemniscatic" case) or a sixth root of unity; so far as the immediate applications of his theory to arithmetic were concerned, Eisenstein was chiefly interested in those two cases and used them to derive the reciprocity laws for fourth and sixth powers. This aspect of his work will not be discussed here.

As before, we put $\tau = \delta v/u$; if W admits the complex multiplier ω, we take $u' = \omega u$, $v' = \omega v$, (u', v') being related to (u, v) by (7). Then we also have $\tau = \delta v'/u'$

and $\omega = a + c\delta\tau$, so that τ is in the field $k = \mathbf{Q}(\omega)$; we have $W = uW_1$, where W_1 is the lattice generated by 1 and τ and is therefore contained in k. Conversely, let the lattice W be such that $v/u \in k$; then we can write $Au^2 + Buv + Cv^2 = 0$ with rational integers A, B, C, and W admits the complex multiplier $\omega = Cv/u$.

§ 7. It will now be shown that, if W admits a complex multiplier, then $e_4^3 e_6^{-2}$ is algebraic over $\mathbf{Q}$; this should be understood to include the "lemniscatic case" $e_6 = 0$, $\omega = i$.

We first observe that an arbitrary lattice W is uniquely characterized by its "invariants" $e_4(W), e_6(W)$. In fact, as we have seen in Chap. IV, § 2, the function $\wp = E_2 - e_2$ satisfies the differential equation

$$y'^2 = 4y^3 - 60 e_4 y - 140 e_6 ;$$

as every solution of this equation is then of the form $y = \wp(x+c)$, and as $\wp$ is the unique solution with a pole at $x=0$, W can be characterized as the lattice of the periods of any solution, or as the lattice of the poles of $\wp$. From this it follows that, if W' is another lattice and t is any element of $\mathbf{C}^\times$, W' coincides with tW if and only if

$$e_4(W') = t^{-4} e_4(W), \quad e_6(W') = t^{-6} e_6(W).$$

Consequently, W' is of the form tW with some $t \in \mathbf{C}^\times$ if and only if the "absolute invariant" $e_4^3 e_6^{-2}$ has the same value for W and for W'.

In particular, let W and W' be as in § 5; let ω be a characteristic root of (7), and assume that it is imaginary; we have $W' = \omega W$ if and only if $e'_4 = \omega^{-4} e_4$, $e'_6 = \omega^{-6} e_6$.

For arbitrary u, v, we have $W' \neq \omega W$ and therefore $e'_4 \neq \omega^{-4} e_4$ or $e'_6 \neq \omega^{-6} e_6$; for instance assume the former inequality and put $f_4 = e'_4 - \omega^{-4} e_4$; this is not 0. From our results in § 5 it follows that f_4 is algebraic over $\mathbf{Q}(e_4, e_6)$, so that it satisfies a non-trivial relation $F(f_4, e_4, e_6) = 0$ with rational coefficients. This can be written as $f_4^m G(f_4, e_4, e_6) = 0$, where G is a polynomial with rational coefficients and not a multiple of f_4; as $f_4 \neq 0$, we still have $G(f_4, e_4, e_6) = 0$ for every lattice W. Consequently, if, for some lattice, f_4 is 0, it must satisfy the non-trivial condition $G(0, e_4, e_6) = 0$. As e_4, e_6 are homogeneous in u, v, of respective degrees $-4, -6$, this must imply that $e_4^3 e_6^{-2}$ is ∞ or algebraic over $\mathbf{Q}$.

§ 8. As we have seen, if W has the complex multiplier ω, it is of the form uW_1, where W_1 is a lattice contained in $k = \mathbf{Q}(\omega)$; clearly, for a given k, all lattices such as W_1 are commensurable with one another, and we can apply to them the results of § 5, combining them with those of § 7. This will be done now.

Let therefore k be an imaginary quadratic field; call $-m$ its discriminant, and Ω the ring of all integers in k; Ω is a lattice, and all lattices in k are commensurable with it. Take a basis $(1, \tau_0)$ for Ω; we may assume that $\text{Im}(\tau_0) > 0$;

as to $\operatorname{Re}(\tau_0)$, it is $\equiv 0$ or $\equiv \frac{1}{2} \bmod 1$ according as m is even or odd. Consequently, if we put $q_0 = \mathbf{e}(\tau_0)$, q_0 is real; it is >0 or <0 according as m is even or odd.

Now formula (36) of Chap. IV, § 11, gives:

(8)
$$\Delta(\Omega) = \Delta(1, \tau_0) = A e_4^3 - B e_6^2 = (2\pi \eta_0^2)^{12},$$

$$\eta_0 = q_0^{1/24} \prod_{n=1}^{\infty} (1 - q_0^n), \quad A = 2^6 \, 3^3 \, 5^3, \quad B = 2^4 \, 3^3 \, 5^2 \, 7^2;$$

$\Delta(\Omega)$ is real and has the sign of q_0.

It will be convenient to introduce the constant

(9)
$$\varpi = 2\pi |\eta_0|^2 = 2\pi |q_0|^{1/12} \prod_{n=1}^{\infty} (1 - q_0^n)^2;$$

this depends solely upon k, i.e. upon m, and we will write ϖ_m for it whenever necessary.

We have $\Delta(\Omega) = \pm \varpi^{12}$. At the same time, it has been shown in § 7 that $e_4^3 e_6^{-2}$ is algebraic over $\mathbf{Q}$. In view of (8), this proves that $e_4 \varpi^{-4}$ and $e_6 \varpi^{-6}$ are algebraic over $\mathbf{Q}$; this amounts to saying that the quantities

$$e_4(\varpi \Omega), \quad e_6(\varpi \Omega)$$

are algebraic over $\mathbf{Q}$.

Let W_1 be a lattice contained in k; as it is commensurable with Ω, ϖW_1 is commensurable with $\varpi \Omega$; the results of § 5 show now that, for all $m \geq 2$, $e_{2m}(\varpi W_1)$ is algebraic over $\mathbf{Q}$. More generally, let W be any lattice with a complex multiplier $\omega \in \Omega$; let u, v be generators for W. As we have seen, the lattice $W_1 = u^{-1} W$ is contained in k; in particular, v/u is in k. We have now:

$$e_{2m}(W) = (\varpi/u)^{2m} e_{2m}(\varpi W_1).$$

Therefore, *if W is a lattice with a complex multiplier $\omega \in \Omega$, the quantities*

$$(u/\varpi)^{2m} e_{2m}(W) \quad (m \geq 2)$$

are all algebraic over $\mathbf{Q}$. In particular, if $e_4(W)$ and $e_6(W)$ are algebraic over $\mathbf{Q}$, so are the numbers $\varpi^{-1} w$ for all $w \in W$.

Chapter VI

Variation II

§ 1. As we have seen in Chap. IV, Eisenstein's method provides an easy access to the derivatives $\partial E_n/\partial v$, $\partial e_n/\partial v$ with respect to v; this is one of its virtues. Using the results in Chap. III, § 5, one can then exchange u with v, and so obtain formulas for the derivatives with respect to u.

At this stage, however, it seems convenient to begin borrowing from Kronecker and start introducing real-analytic functions into the theory, besides the complex-analytic ones. In the present chapter this will be more formal than substantial; the functions chiefly to be considered here will be polynomials in the complex-conjugates $\bar{x}, \bar{u}, \bar{v}$ of x, u, v, whose coefficients will be complex-analytic functions of x, u, v. The advantage of this procedure will be to lead to more manageable formulas, for reasons to be explained presently.

As is customary in dealing with real-analytic functions, we treat formally $x, u, v, \bar{x}, \bar{u}, \bar{v}$ as independent variables in writing partial derivatives; in applying $\partial/\partial x$, $\partial/\partial u$, $\partial/\partial v$, to functions which are polynomials or rational functions with respect to $\bar{x}, \bar{u}, \bar{v}$, the latter variables are to be treated as constants.

Now we introduce the differential operator:

(1) $$\mathscr{D} = \bar{x}\frac{\partial}{\partial x} + \bar{u}\frac{\partial}{\partial u} + \bar{v}\frac{\partial}{\partial v}.$$

Take any homogeneous function f of degree $-n$ in x, u, v (e.g. E_n); we have

(2) $$x\frac{\partial f}{\partial x} + u\frac{\partial f}{\partial u} + v\frac{\partial f}{\partial v} = -nf.$$

Consequently, one can express $\partial f/\partial u$ and $\mathscr{D}f$ in terms of f, $\partial f/\partial x$, $\partial f/\partial v$, but also (since $u\bar{v} - v\bar{u}$ is not 0) $\partial f/\partial u$, $\partial f/\partial v$ in terms of f, $\partial f/\partial x$ and $\mathscr{D}f$. The advantage of the operator $\mathscr{D}$ is of course that it is invariant under the group $GL(3,\mathbf{R})$, i.e. under any real linear substitution on x, u, v. In this chapter we shall use it systematically, instead of $\partial/\partial v$.

§ 2. Notations will be the same as before; in particular, as in Chap. III, § 1, we write

(3) $\quad u\bar{v} - v\bar{u} = \delta u\bar{u}(\bar{\tau} - \tau) = -2\pi i \delta A, \quad A > 0;$

clearly we have $\mathscr{D}A = 0$.

As u, v make up a basis of $\mathbf{C}$ over $\mathbf{R}$, we can write x in terms of that basis as $x = \alpha u + \beta v$ with real α, β; for α, β we also write $\alpha(x), \beta(x)$ when necessary. We have $\bar{x} = \alpha \bar{u} + \beta \bar{v}$, and therefore:

$$\beta(x) = \frac{\bar{u}x - u\bar{x}}{2\pi i \delta A}, \quad \mathscr{D}\beta = 0.$$

Next, we introduce, beside E_1, the function

(4) $\quad E_1^*(x) = E_1(x) + \dfrac{2\pi i \delta}{u}\beta(x).$

It is not complex-analytic in x, but it compensates for this "defect" by being periodic with the period-lattice W, as formula (5) of Chap. III, § 4, shows at once.

Furthermore, formula (7) of Chap. III, § 5, shows that, under any change of basis for the lattice W, $E_1(x; u, v)$ is changed into a function differing from it only by a linear function; then E_1^* can be uniquely characterized as the periodic function, with the period-lattice W, which differs from any one of these functions $E_1(x; u', v')$ only by a real-linear function. Consequently, E_1^* depends only upon the lattice W and not upon the choice of the generators u, v, and we may write it as $E_1^*(x; W)$. More generally, if W' is a sublattice of W, and if R is any set of representatives for W/W' in W, formula (7) of Chap. III, § 5, gives, after a trivial calculation:

(5) $\quad \sum_{r \in R} E_1^*(x + r; W') = E_1^*(x; W).$

Similarly, beside E_2, e_2, we introduce

(6) $\quad E_2^*(x) = -\dfrac{\partial E_1^*}{\partial x} = E_2 - \dfrac{\bar{u}}{Au}, \quad e_2^* = e_2 - \dfrac{\bar{u}}{Au};$

clearly these, too, depend only upon W, and E_2^* satisfies a formula similar to (5). As E_2^* differs from E_2 only by a constant, further differentiation with respect to x will only produce the functions E_n once more, with $n \geq 3$; for greater formal symmetry in our formulas we shall write $E_n^* = E_n$ for all $n \geq 3$.

§ 3. As E_1^* is odd and periodic, it must (just as E_3; cf. Chap. IV, § 11) take the value 0 at all points of $\frac{1}{2}W$, not in W. Moreover, since it differs from E_1 only by a linear function, it satisfies the same addition formula as E_1, i.e. (10) of Chap. IV, §2; in the latter formula, we can also substitute $E_2 - e_2$ for E_2 in order to apply to it the results of Chap. V, §4. Now define a function f on $\mathbf{Q}W$ by putting $f(w) = 0$ for every w in W, and $f(x) = E_1^*(x)$ for every x in $\mathbf{Q}W$, not in W. Then the addition formula, combined with §4 of Chap. V, shows that $f(x+x') - f(x) - f(x')$ is algebraic over $\mathbf{Q}(e_4, e_6)$ for all x, x' in $\mathbf{Q}W$. As f is 0 on $\tfrac{1}{2}W$, this obviously implies that the values of f, and therefore also those of E_1^*, are algebraic over $\mathbf{Q}(e_4, e_6)$ at all points of $\mathbf{Q}W$, not in W.

Assume now that W has a complex multiplier ω; put $W' = \omega W$, and take the formula for E_2^* obtained from (5) by differentiation. As E_2^* is homogeneous of degree -2 in x, u, v, and of degree 0 in $\bar{u}, \bar{v}$, we can write this as follows:

$$\omega^{-2} \sum_{r \in R} E_2^*\left(\frac{x+r}{\omega}; W\right) = E_2^*(x; W).$$

Subtract x^{-2} from both sides and put $x = 0$. Taking into account that the number of terms in the left-hand side is $N = \omega \bar{\omega}$, we get

(7) $$\sum_{r \in R'} (E_2^*(r/\omega) - e_2^*) = \omega(\omega - \bar{\omega}) e_2^*,$$

where R' has the usual meaning. Now observe that $E_2^* - e_2^*$ is the same as $E_2 - e_2$; moreover, as the trace $\omega + \bar{\omega}$ is a rational integer, $\bar{\omega} W$ is contained in W, and so is $Nr/\omega = \bar{\omega} r$. Applying now the results of Chap. V, §4, to $E_2 - e_2$ and r/ω, we see that the left-hand side of (7) is algebraic over $\mathbf{Q}(e_4, e_6)$; in particular, it is algebraic over $\mathbf{Q}$ if e_4 and e_6 are so. As e_2^*, like E_2^*, is homogeneous of degree -2 in u, v, and of degree 0 in $\bar{u}, \bar{v}$, we can now apply the final results of Chap. V, and conclude that $(u/\bar{\omega})^2 e_2^*(W)$ is algebraic over $\mathbf{Q}$.

§4. Our purpose is now to study the functions obtained from E_1^* by repeated application of the operators $\partial/\partial x$ and $\mathscr{D}$; clearly these operators commute with each other. For convenience, we shall write $\partial_x = \partial/\partial x$. Whenever a, b are two integers such that $b > a \geq 0$, we shall write:

(8) $$E_{a,b}(x) = \sum_{w \in W}{}_e (\bar{x}+\bar{w})^a (x+w)^{-b} = \frac{(-1)^{b-1}}{(b-1)!} \mathscr{D}^a \partial_x^{b-a-1} E_1,$$

$$E_{a,b}^*(x) = \frac{(-1)^{b-1}}{(b-1)!} \mathscr{D}^a \partial_x^{b-a-1} E_1^*.$$

For $b \geq a+3$, the series for $E_{a,b}$ is absolutely convergent, and $E_{a,b}^*$ is the same as $E_{a,b}$; otherwise the difference $E_{a,b}^* - E_{a,b}$ is easily calculated from (4) and (6) of §2. For all $n \geq 1$, $E_{0,n}$ coincides with E_n, and $E_{0,n}^*$ with E_n^*.

As E_1^* is homogeneous of degree -1 in x, u, v (and of degree 0 in $\bar{x}, \bar{u}, \bar{v}$), we can apply to it our formula (2) of §1. On the other hand, $\partial E_1/\partial v$ is given by (9) of Chap. IV, § 2. Combining those results with the definitions of E_1^* and E_2^*, one gets, after a trivial calculation:

(9) $\qquad E_{1,2}^* = -\mathscr{D}(E_1^*) = A(E_3^* - E_1^* E_2^*).$

Differentiating this $n-1$ times with respect to x, we get:

(10) $\qquad E_{1,n+1}^* = -\dfrac{1}{n}\mathscr{D}(E_n^*) = A\left(\dfrac{n+1}{2}E_{n+2}^* - \dfrac{1}{2}\sum_{h=0}^{n} E_{h+1}^* E_{n-h+1}^*\right).$

Using this formula and induction on a, one finds that $E_{a,b}^*$ is given, for all a, b such that $b > a \geq 0$, by a formula of the form

(11) $\qquad E_{a,b}^* = \dfrac{(A/2)^a}{(b-1)(b-2)\ldots(b-a)} P_{a,b}(E_1^*, E_2^*, E_3, \ldots, E_{a+b}),$

where $P_{a,b}$ is a polynomial in $a+b$ indeterminates, of degree $a+1$, with rational integral coefficients. Moreover, A and $E_{a,b}^*$ are homogeneous, of respective degrees 1 and $-b$ in x, u, v and of degrees 1 and a in $\bar{x}, \bar{u}, \bar{v}$, while each E_n^* is homogeneous of degree $-n$ in x, u, v, and of degree 0 in $\bar{x}, \bar{u}, \bar{v}$; consequently $P_{a,b}$ must be "isobaric" of "weight" $a+b$.

§ 5. Similarly, for $b > a \geq 0$ we define $e_{a,b}$, $e_{a,b}^*$ as being the values of $E_{a,b} - \bar{x}^a x^{-b}$, $E_{a,b}^* - \bar{x}^a x^{-b}$, for $x=0$; we have

$$e_{a,b} = \dfrac{(-1)^a}{(b-1)(b-2)\ldots(b-a)}\mathscr{D}^a(e_{b-a}),$$

and a similar formula for $e_{a,b}^*$; of course these are 0 unless $b-a$ is even.

Starting now from formula (2) of Chap. V, § 1, and proceeding just as above in § 4, we get

$$e_{1,2m+1}^* = -\dfrac{1}{2m}\mathscr{D}(e_{2m}^*) = A\left(\dfrac{2m+3}{2}e_{2m+2} - \dfrac{1}{2}\sum_{r=1}^{m} e_{2r}^* e_{2m-2r+2}^*\right),$$

and from this, by induction on a, a formula

$$e_{a,b}^* = \dfrac{(A/2)^a}{(b-1)\ldots(b-a)} Q_{a,b}(e_2^*, e_4, \ldots, e_{a+b}),$$

where $Q_{a,b}$ is an "isobaric" polynomial of degree $a+1$ and weight $a+b$ with rational integral coefficients; of course it is 0 unless $a+b$ is even.

§ 6. Assume that W is a lattice with a complex multiplier ω; define k and ϖ as in Chap. V, § 8; formula (3), § 2, shows that $\pi i A/u\bar{u}$ is in k.

Combining now §5 with the final results of Chap. V, we see that $A^{-a}(u/\varpi)^{a+b}e^*_{a,b}$ (or, what amounts to the same, $\pi^a\varpi^{-a-b}u^b\bar{u}^{-a}e^*_{a,b}$) is algebraic over $\mathbf{Q}$.

Similarly, combining the results of §§3—4 with those of Chap. V, we find that the values of the function

$$\pi^a\varpi^{-a-b}u^b\bar{u}^{-a}E^*_{a,b}(x)$$

at the points x of $\mathbf{Q}W$, not in W, are all algebraic over $\mathbf{Q}$.

In particular, if the lattice W is contained in the imaginary quadratic field k, u and $\bar{u}$ themselves are in that field; therefore, when that is so, the quantities

$$\pi^a\varpi^{-a-b}e^*_{a,b}, \quad \pi^a\varpi^{-a-b}E^*_{a,b}(x),$$

for all x in $\mathbf{Q}W=k$ and not in W, are all algebraic over $\mathbf{Q}$, whenever $b>a\geq 0$.

We can now compare this with Damerell's theorem, quoted in Chap. V, §1. The latter refers to values of Hecke L-functions

$$L(s)=\sum_{\mathfrak{a}}\chi(\mathfrak{a})N\mathfrak{a}^{-s};$$

here χ is a Hecke character of ideals in k; this means that there is an ideal $\mathfrak{f}$ in k (the "conductor" of χ) such that $\chi(\mathfrak{a})$ is 0 if and only if $\mathfrak{a}$ is not prime to $\mathfrak{f}$ and that $\chi((\alpha))=\alpha^e\bar{\alpha}^f$ whenever α is an integer in k and $\alpha\equiv 1\bmod\mathfrak{f}$; here e,f are two rational integers, and one may assume, without loss of generality, that $e=0$, $f>0$; if e and f were both 0, χ would be an "ordinary" character (of finite order) and not a genuine Hecke character. The series for $L(s)$ is absolutely convergent in the half-plane $\operatorname{Re}(s)>1+\dfrac{f}{2}$.

Take $s=b$, where b is an integer in the half-plane of absolute convergence, and put $a=f-b$; assume also that $a\geq 0$, i.e. $b\leq f$. As b is $>1+\dfrac{f}{2}$, we have $b\geq a+3$. It is now easily seen that $L(b)$ is a linear combination, with algebraic coefficients, of finitely many quantities of the form $e_{a,b}(W)$, $E_{a,b}(x;W)$, where W is one of finitely many lattices contained in k, and $x\in k$. Therefore $\pi^a\varpi^{-a-b}L(b)$ is then algebraic over $\mathbf{Q}$.

Damerell's full result is that this last statement remains valid provided $b\geq\dfrac{1}{2}+\dfrac{f}{2}$, i.e. provided $b\geq a+1$; using the functional equation for L, one can then extend it to all the cases $1\leq b\leq f$. In order to complete the proof of Damerell's theorem, we have to show that, when $b-a$ is 1 or 2, we have the same relation between $L(b)$ and the quantities $e^*_{a,b}$, $E^*_{a,b}(x)$ as we found above in the case of absolute convergence. As the definition of $L(b)$ for $b\leq 1+\dfrac{f}{2}$

requires the analytic continuation of the series for $L(s)$, which is one of the highlights of Kronecker's treatment of our topic, this is best postponed to a later chapter.

Part II
KRONECKER

Chapter VII
Prelude to Kronecker

§ 1. Kronecker was born in 1823, the same year as Eisenstein; they were students in Berlin at the same time. In 1847 Kronecker had to leave Berlin to take care of the business interests of his family; by the time he came back to settle there permanently, Eisenstein was dead.

The first signs of an awakening interest in elliptic functions, on the part of Kronecker, appear in 1853 (*Werke* IV, p. 11); there he merely mentions the lemniscatic case as providing the generalization to the Gaussian field $\mathbf{Q}(i)$ of his theorem on the abelian extensions of $\mathbf{Q}$. Undoubtedly he must then have studied, besides Abel, the work of Eisenstein on the division of the lemniscate; but this (even Eisenstein's great paper of 1850) was based on Abel's formulas and notations and bore no close relation to the *Genaue Untersuchung* of 1847 which has been described above in Chapters I to IV.

In 1856 we find Kronecker extending his investigations to the general case of elliptic functions with complex multiplication (*Werke* IV, p. 179, and V, p. 419); here he makes exclusive use of Jacobi's notations, to which he was to remain faithful ever after.

In 1863 (*Werke* IV, p. 222), under the influence of Dirichlet's work, he introduces new functions

$$\sum_{\mu,\nu}{}' \frac{e^{2\pi i(r\mu+s\nu)}}{(a\mu^2+2b\mu\nu+c\nu^2)^{1+\rho}}$$

and their limits for $\rho=0$; in the notation we have used in earlier chapters, this can be written as

(1) $$\sum_{w\in W}{}' \chi(w)(w\bar{w})^{-1-\rho}$$

where χ is a character of the additive group W. It is in this paper that Kronecker states for the first time a partial result on his "limit-formula" and deduces from it a solution of Pell's equation (i.e., the determination of a unit in a real quadratic field) by elliptic functions.

§ 2. Twenty years later, after some desultory publications on the same subject, Kronecker finally made up his mind to work out his ideas systematically in a series of memoirs, to be published by the Berlin Academy. They appeared, under the title *Zur Theorie der elliptischen Funktionen*, in 1883, 1885, 1886, 1889 and 1890; in 1891, the last year of Kronecker's life, the title was changed (without any compelling reason) into *Die Legendre'sche Relation*. Kronecker's chief concern in these papers is with various series which (in our notation) are all of the form

(2) $$\sum_{w \in W} \chi(w)(\bar{x}+\bar{w})^a |x+w|^{-2s},$$

where a is an integer ≥ 0, and with their behavior for $x=0$ and for $s=1+\dfrac{a}{2}$ (when they cease to be convergent).

Kronecker's mind was too quick and too restless, however, to allow him to settle down to the systematic exposition of any subject; in his younger years Kummer and Dirichlet had already warned him against this defect. His students were used to seeing him turn aside from the current topic and report on whatever had occurred to him the night before. In the Academy memoirs we find him frequently shifting his ground or improving upon earlier proofs and results. His contribution of 1886 (*Werke* IV, pp. 389—470), although ostensibly part of the main series, bears no relation to it and is devoted to a purely algebraic and number-theoretic investigation of the multiplication and division formulas for elliptic functions; this is where he proves the famous congruences which play the main role in the arithmetical theory of complex multiplication.

Eisenstein had obviously prided himself on the wholly elementary nature of his function-theoretic methods. In contrast with this, Kronecker has at his disposal a whole arsenal of powerful tools: "Poisson summation" (actually due to Cauchy); Cauchy's theory of residues; Dirichlet's theory of Fourier series; and, most important of all, Dirichlet's transformation formula (essentially our Mellin transform)

(3) $$\Gamma(s) \sum \frac{a}{A^s} = \int_0^\infty (\sum a e^{-At}) t^{s-1} dt,$$

which (following Dirichlet) he prefers to write

$$\Gamma(1+\rho) \sum \frac{a}{A^{1+\rho}} = \int_0^1 (\sum a z^A)\left(\log \frac{1}{z}\right)^\rho d(\log z).$$

To this a modern analyst would have little to add, except for the concept of analytic continuation (which Kronecker knew but chose to avoid) and for the freer use of Fourier series which is allowed by the theory of distributions.

Clearly Kronecker's series were a natural generalization of Eisenstein's. In introducing the continuous parameter ρ (or, in Riemann's notation and ours, s), he was following Dirichlet; precedents can be found for the appearance of the character χ. In dealing with his series when they fail to converge, Kronecker also makes frequent use of "Eisenstein summation". But, until 1891, he makes no mention of Eisenstein's paper of 1847; only then, while writing his last memoir for the Berlin academy, did he realize how close he was to the companion of his youth. We are left to guess what his personal feelings may have been on making that discovery. Had he lived to write his lecture on Eisenstein as he had promised to Cantor (cf. above, Chap. I), he would perhaps have told us more.

§ 3. Just as Eisenstein before him (cf. above, Chap. II), Kronecker found that a study of double series such as (2) requires a preliminary study of the corresponding simple series; moreover, in both cases, the same kind of machinery is involved. Eventually Kronecker devoted to such simple series considerable portions of two separate memoirs (*Werke* V, pp. 267—294 and 327—342); in this, as he acknowledged (ibid., p. 330), he had been largely anticipated by Lipschitz. There are obvious relations between those series, Dirichlet's L-functions, and Riemann's (or rather Euler's) zeta-function. It is therefore not surprising that several authors should have dealt with this topic in the course of the last century, Lipschitz as early as 1857, Hurwitz in 1882, and Lerch a little later under the influence of Kronecker's work. A short bibliography will be found at the end of this chapter.

To complicate matters further, Poisson summation, applied to these series (for imaginary values of the argument) leads to Bessel functions, whose theory was well advanced in Kronecker's time; but, even quite recently, some of the authors who dealt with this topic have failed either to notice or to point out the occurrence of Bessel functions, and have contented themselves with direct proofs for the few elementary properties which they needed.

The attempt will not be made here to disentangle all these threads. In this chapter we will merely treat the simple series as a preparation for our description of Kronecker's work on the double series of type (2).

§ 4. In this chapter we denote by χ a character of the group **Z**; we will usually write it as

$$\mu \mapsto \chi(\mu) = \mathbf{e}(-\mu y)$$

with $y \in \mathbf{R}$; whenever convenient, one may assume $0 \leq y < 1$. We consider the series

(4) $$S_a(x, y, s) = \sum_{\mu}{}^* (\bar{x} + \mu)^a |x + \mu|^{-2s} \mathbf{e}(-\mu y),$$

where a is an integer ≥ 0, y is real, x and s are complex, and where $\sum^*$ denotes the sum taken over all integers $\mu \neq -x$ (i.e. over all integers, unless x is an integer). The series is absolutely convergent if and only if $\mathrm{Re}(s) > \dfrac{a+1}{2}$; but, as Kronecker (and, before him, Lipschitz) observed, it is convergent in the ordinary sense provided $\chi \neq 1$ and $\operatorname{Re}(s) > a/2$. In fact, if we write the general term of (4) as $f(\mu)\chi(\mu)$ and put

$$\sigma_n = \sum_{\mu=0}^{n} \chi(\mu),$$

we get, by Abel's method of "partial summation":

(5) $$\sum_{\mu=1}^{N} f(\mu)\chi(\mu) = f(N)\sigma_N - f(1) + \sum_{n=1}^{N-1} [f(n) - f(n+1)]\sigma_n.$$

As $\chi \neq 1$, $|\sigma_n|$ is bounded for all $n > 0$. On the other hand, for n large, $n^{2s-a} f(n)$ and $n^{2s-a} f(n+1)$ can be expanded into power-series in n^{-1}, both with the constant term 1; therefore

$$|n^{2s-a+1}[f(n) - f(n+1)]|$$

is bounded, and, for $N \to +\infty$, the last sum on the right-hand side of (5) becomes an absolutely convergent series provided $\operatorname{Re}(s) > a/2$. As the same argument applies to the terms of (4) for $\mu < 0$, this proves the convergence of that series.

A case of particular interest is that of the series in the formula

(6) $$\sum_{-\infty}^{+\infty} \frac{\mathbf{e}(-\mu y)}{x+\mu} = 2\pi i \frac{\mathbf{e}(xy)}{\mathbf{e}(x)-1} \quad (0 < y < 1),$$

valid for any x in $\mathbb{C}$ and not in $\mathbb{Z}$; as Kronecker observes, this is an immediate consequence of Dirichlet's theorem on the expansion of a periodic function into a Fourier series, applied to the discontinuous function of y of period 1, equal to $\mathbf{e}(xy)$ for $0 < y < 1$ and to $\frac{1}{2}[1 + \mathbf{e}(x)]$ for $y = 0$.

Subtracting x^{-1} from both sides in (6), and then taking $x = 0$, one gets

(7) $$\sum_{-\infty}^{+\infty}{}' \mu^{-1} \mathbf{e}(-\mu y) = \pi i(2y - 1) \quad (0 < y < 1),$$

which could also be verified in the same manner as above. More generally, expanding both sides of (6) into power-series at $x = 0$, one gets coefficients which are essentially the Bernoulli polynomials in y; this extends the fact that we obtained the Bernoulli numbers, in Chap. II, § 7, by expanding $\varepsilon_1(x)$ at $x = 0$. If for instance we take the coefficient of x in both sides in (6), we get

(8) $$\sum_{-\infty}^{+\infty}{}' \mu^{-2} \mathbf{e}(-\mu y) = 2\pi^2(y^2 - y + \tfrac{1}{6}) \quad (0 \leq y \leq 1);$$

here the series is absolutely convergent, so that, by continuity, the formula remains valid for $y=0$ and $y=1$.

Apart from the above cases of absolute and of "ordinary" convergence, there are cases where $\sum_e$ (as defined in Chap. II, § 1) converges. This is so, for instance, when x is real, $\chi = 1$ and $a = 2s = 1$; we have then

$$\sum_e \operatorname{sgn}(x+\mu) = 2[x] + 1$$

provided x is not an integer; as usual, $[x]$ denotes the integer n such that $n < x < n+1$.

§ 5. While Kronecker experimented with various methods of summation (including Eisenstein's) for the series (2) and (4), perhaps his favorite was the one inspired by the work of Dirichlet, which depends upon the use of the parameter s. Let $S(s)$ be any one of these series; it is absolutely convergent in a half-plane, $\operatorname{Re}(s) > \frac{a}{2} + 1$ in case (2) and $\operatorname{Re}(s) > \frac{a+1}{2}$ in case (4). Let s_0 be on the edge of that half-plane. Then $S(s_0 + \rho)$ is absolutely convergent for ρ real and >0. If this has a limit for $\rho \to 0$, the limit will be called the value of $S(s_0)$ under "Kronecker summation". Kronecker also considered variants of this, where absolute convergence is replaced by "ordinary" or "Eisenstein" convergence.

Kronecker summation may of course be regarded as a special case of analytic continuation. In fact, as we shall see, all the series (2) and (4) can be continued as meromorphic functions of s in the whole s-plane; this was established (following Riemann's break-through in his paper of 1859 on the zeta-function) by Kronecker's contemporaries, Lipschitz, Hurwitz and Lerch. Indeed the proof for it was virtually contained in some of Kronecker's calculations (*Werke* IV, pp. 486—487), but he never mentions analytic continuation; perhaps his increasingly cool relations with Weierstrass had made the concept distasteful to him. We will of course not avoid it in our exposition. To attribute to $S(s)$ the value obtained by analytic continuation from the half-plane of absolute convergence might well be called "Hecke summation", in view of the extensive use made by Hecke of this method.

§ 6. Since we plan to give here only those results about (4) which will be needed in Chap. VIII, it will be convenient to deal separately with the case where x is real and the case where it is not. When x is real, (4) can be rewritten as

$$\sum{}^* \operatorname{sgn}(x+\mu)^a |x+\mu|^{a-2s} \mathbf{e}(-\mu y),$$

so that it is enough to consider the cases $a = 0$ and $a = 1$. The investigation of such series is equivalent to that of "Lerch's series"

(9) $$\sum_{n=0}^{+\infty} \chi(n)(x+n)^{-s}$$

for $x > 0$; clearly Dirichlet's L-series are finite linear combinations of such series with χ of finite order and x rational and ≤ 1. One should also keep in mind that the "one-sided" series (9) bears more or less the same relationship to the gamma-function as Eisenstein's series $\varepsilon_n(x)$ bear to the Euler product for the sine (cf. Chap. II, § 6), so that a theory of the gamma-function could be built up on the basis of this analogy. For the sake of brevity, we will take that theory for granted.

§ 7. Consequently, with the notations explained above, we write:

$$(10) \qquad S_a(x,y,s) = \sum_{\mu}{}^{*} (x+\mu)^a |x+\mu|^{-2s} \mathbf{e}(-\mu y),$$

where x and y are real, and $a = 0$ or 1. Applying (3) formally, we get:

$$(11) \qquad \Gamma(s) S_a(x,y,s) = \int_0^{+\infty} \sum{}^{*} \exp[-t(x+\mu)^2 - 2\pi i \mu y](x+\mu)^a \, t^{s-1} \, dt.$$

On both sides, replace y by 0, $(x+\mu)^a$ by $|x+\mu|^a$ and s by $\mathrm{Re}(s)$; then both series become series with positive terms, and the left-hand side converges provided $\mathrm{Re}(s) > \frac{a+1}{2}$; therefore, if that is so, termwise integration is permissible in the right-hand side of (11), and both sides are equal. Now take any $T > 0$ and cut up the integral in the right-hand side into an integral I_0 on the interval $0 < t \leq T$ and an integral I_∞ on $t > T$. The latter can be estimated by observing that, if M is an integer $\geq |x|$, we have

$$\exp[-t(x+\mu)^2] \cdot |x+\mu|^a \leq \exp(-tn^2) \cdot (n+2M)$$

for $\mu > M$, $n = \mu - M$, and also for $\mu < -M$, $n = -\mu - M$. Using this, it is easy to verify that I_∞ is absolutely convergent for all values of s, uniformly over every bounded domain in the s-plane; it is therefore an entire function of s.

As to I_0, one applies the well-known transformation formula for the theta-series:

$$(12) \qquad \sum_\mu \exp[-t(x+\mu)^2 - 2\pi i \mu y](x+\mu)^a$$
$$= i^{-a} \mathbf{e}(xy) \left(\frac{\pi}{t}\right)^{a+\frac{1}{2}} \sum_v \exp\left[-\frac{\pi^2}{t}(y+v)^2 + 2\pi i v x\right](y+v)^a.$$

This is easily obtained by Poisson summation, or, what amounts to the same, by observing that the right-hand side is periodic in y with the period 1, and expanding it into its Fourier series. Alternatively, one can obtain (12) by Poisson summation for $a = 0$, and then differentiate with respect to x in order to obtain the formula for $a = 1$.

Now apply (12) to the integrand in I_0 and make the change of variable $t=\pi^2/u$; one gets an integral similar to I_∞, with possibly two additional terms corresponding respectively to the term $\mu = -x$ in the left-hand side of (12) if x is an integer, and to $\nu = -y$ in the right-hand side if y is an integer; as we have assumed $\mathrm{Re}(s) > \frac{a+1}{2}$, those terms (which occur only if $a=0$) can be integrated at once. The final result is as follows. Instead of I_∞, write more explicitly $I_\infty(T,x,y,a,s)$, and put

(13) $\quad I'_\infty = I_\infty(\pi^2/T, y, -x, a, a-s+\tfrac{1}{2})$,

which is also an entire function of s, since this was the case for I_∞. Put $\varepsilon = 1$ if x is an integer and $a=0$, and $\varepsilon = 0$ otherwise; put $\varepsilon' = 1$ if y is an integer and $a=0$, and $\varepsilon' = 0$ otherwise. Then:

(14) $\quad \Gamma(s)S_a(x,y,s) = I_\infty + i^{-a}\pi^{2s-a-\frac{1}{2}}\mathbf{e}(xy)I'_\infty - \varepsilon\mathbf{e}(xy)\dfrac{T^s}{s} + \varepsilon'\sqrt{\pi}\dfrac{T^{s-\frac{1}{2}}}{s-\frac{1}{2}}$.

This shows that the left-hand side can be continued to the whole s-plane as a meromorphic function, with possible poles at $s=0$ and $s=\tfrac{1}{2}$. Defining $S_a(x,y,s)$ for all s by means of (14), and taking $T=\pi$, one verifies at once the functional equation:

(15) $\quad \Gamma(s)S_a(x,y,s) = i^{-a}\pi^{2s-a-\frac{1}{2}}\mathbf{e}(xy)\Gamma(a-s+\tfrac{1}{2})S_a(y,-x,a-s+\tfrac{1}{2})$.

For rational values of x and y, this is in substance equivalent to the functional equations of Dirichlet's L-functions.

§ 8. It was shown in § 4 that the right-hand side of (10) converges in the sense of "ordinary" convergence if y is not an integer and if $\mathrm{Re}(s) > a/2$; the same proof can easily be adapted to show that the sum is then holomorphic in s and therefore still equal to $S_a(x,y,s)$; this need of course not be true otherwise, even if that series can be summed in the sense of ordinary convergence or of Eisenstein summation. We shall now examine a few cases of special interest.

We begin with $S_1(x,0,\tfrac{1}{2})$ when x is not an integer; formally, this corresponds to the series $\sum \mathrm{sgn}(x+\mu)$, for which Eisenstein summation would give the value $2[x]+1$ (cf. § 4). However, (15) gives

$$S_1(x,0,\tfrac{1}{2}) = \frac{1}{\pi i} S_1(0,-x,1).$$

As we have just observed, the right-hand side is within the domain of "ordinary" convergence; therefore its value is as given by (7). Thus we get:

(16) $\quad S_1(x,0,\tfrac{1}{2}) = 2\langle -x\rangle - 1 = 1 - 2\langle x\rangle$,

where we have put, as usual, $\langle x\rangle = x - [x]$ (this is the "fractional part" of x).

As $\Gamma(s)$ has a pole at $s=0$, (14) shows that $S_a(x,y,0)$ is 0 unless $a=0$ and x is an integer; we shall now calculate $\partial S_0(x,0,s)/\partial s$ for $s=0$. As $\Gamma(s)$ can be written as $s^{-1}\Gamma(1+s)$, (15) gives, at $s=0$, the expansion:

$$S_0(x,0,s) = s\frac{\pi^{2s-\frac{1}{2}}\Gamma(\frac{1}{2}-s)}{\Gamma(1+s)}S_0(0,-x,\tfrac{1}{2}-s) = sS_0(0,-x,\tfrac{1}{2}) + \cdots.$$

As above, the value of $S_0(0,-x,\frac{1}{2})$, when x is not an integer, is given by the convergent series:

$$S_0(0,-x,\tfrac{1}{2}) = \sum_{-\infty}^{+\infty}{}' \frac{\mathbf{e}(\mu x)}{|\mu|}.$$

This happens to be the Fourier series for $\log|2\sin\pi x|^{-2}$, as one could verify directly; it is perhaps more interesting to proceed as follows. Put $f(x) = S_0(0,-x,\frac{1}{2})$; obviously the above series gives $f(\frac{1}{2}) = -2\log 2$. Within the domain of absolute convergence, the series for $S_0(x,0,s)$ can be differentiated term by term; this gives

$$\frac{\partial}{\partial x}S_0(x,0,s) = -2sS_1(x,0,s+1),$$

which, by analytic continuation, must remain valid for all s. Differentiating this with respect to s, and taking $s=0$, we get $df/dx = -2S_1(x,0,1)$. Formally, the series for $S_1(x,0,1)$ is $\sum(x+\mu)^{-1}$, for which Eisenstein summation gives the value $\pi\cot\pi x$. To show that this is also the value of $S_1(x,0,1)$, we observe that they are both odd functions of x; more generally, $S_a(x,0,s)$ is an even or an odd function of x according as $a=0$ or $a=1$. On the other hand, by differentiating the series for $S_1(x,0,s)$ term by term within the domain of absolute convergence, and then using analytic continuation, one finds, just as above for $S_0(x,0,s)$, that $dS_1(x,0,1)/dx$ is $-S_0(x,0,1)$, this being given by the absolutely convergent series $-\sum(x+\mu)^{-2}$. Thus $S_1(x,0,1)$ and $\pi\cot\pi x$ differ only by a constant; as they are odd functions, they coincide. Therefore $f(x)$ differs from $\log|2\sin\pi x|^{-2}$ only by a constant; as they have the same value for $x=\frac{1}{2}$, and as they are both periodic of period 1, they coincide whenever x is not an integer.

§9. Now we wish to calculate $\partial S_1(x,0,s)/\partial s$ for $s=\frac{1}{2}$. Put:

(17) $$H(x,s) = \sum_{n=0}^{+\infty}(x+n)^{-s}$$

for $x>0$, $\mathrm{Re}(s)>1$. We have:

(18) $$H(x+1,s) = H(x,s) - x^{-s},$$

(19) $$\frac{\partial H(x,s)}{\partial x} = -sH(x,s+1).$$

On the other hand, we have, for $\mathrm{Re}(s)>1$ and $0<x\leq 1$:

(20) $\quad H(x,s) = \dfrac{1}{2} S_0\!\left(x,0,\dfrac{s}{2}\right) + \dfrac{1}{2} S_1\!\left(x,0,\dfrac{s+1}{2}\right).$

Together with (18) and (14), this shows that $H(x,s)$, for all $x>0$, can be continued to a meromorphic function in the whole s-plane, with only one pole at $s=1$ and the residue 1 at that pole; this is sometimes known as "Hurwitz' function"; it will still be denoted by $H(x,s)$. As usual, we put

$$\zeta(s) = H(1,s) = \dfrac{1}{2} S_0\!\left(0,0,\dfrac{s}{2}\right),$$

so that (15) provides the well-known functional equation for ζ.

For $s=0$, (14) gives $H(1,0)=\zeta(0)=-\tfrac{1}{2}$; for $s=0$, $0<x<1$, (20), together with (14) and (16), gives

$$H(x,0) = \tfrac{1}{2} - \langle x \rangle.$$

Using (18), which remains valid for all s by analytic continuation, we get

(21) $\quad H(x,0) = \tfrac{1}{2} - x \quad (x>0).$

For any $N\geq 0$, $x>0$ and $\mathrm{Re}(s)>1$, we have

$$H(x,s) - \sum_{i=0}^{N} \binom{s+i-1}{i} \zeta(s+i)(-x)^i$$

$$= x^{-s} + \sum_{n=1}^{+\infty} n^{-s}\left[\left(1+\dfrac{x}{n}\right)^{-s} - \sum_{i=0}^{N} \binom{s+i-1}{i}\left(\dfrac{-x}{n}\right)^i\right];$$

as the right-hand side is absolutely convergent for $\mathrm{Re}(s)>-N$, analytic continuation shows that it remains valid in that half-plane; it may be viewed as providing the analytic continuation for $H(x,s)$, once $\zeta(s)$ has been so continued. Now put $N=1$ in this formula; then differentiate it (term by term) once with respect to s and twice with respect to x, and put $F(x)=(\partial H/\partial s)_{s=0}$; we get, for $s=0$:

$$\dfrac{d^2 F}{dx^2} = \sum_{n=0}^{+\infty} (x+n)^{-2} = H(x,2) > 0.$$

On the other hand, differentiate (18) with respect to s, and put $s=0$; this gives:

$$F(x+1) - F(x) = \log x.$$

It is well known that this, together with the positivity of d^2F/dx^2, characterizes the function $\log \Gamma(x)$ up to an additive constant[8]. Therefore we have $F(x) = \log C\Gamma(x)$, for a suitable value of the constant C.

Now, changing x into $1-x$ in (20), for $0 < x < 1$, and adding this to (20), we get, since S_0 is even in x, S_1 is odd, and both are periodic:

$$(22) \qquad S_0\left(x, 0, \frac{s}{2}\right) = H(x,s) + H(1-x, s),$$

which of course can also be verified directly. Differentiating with respect to s for $s = 0$, we get, in view of the results in § 8:

$$\log|2\sin \pi x|^{-1} = \log[C^2 \Gamma(x) \Gamma(1-x)] \qquad (0 < x < 1).$$

Here it is enough to know that $\Gamma(\tfrac{1}{2}) = \sqrt{\pi}$ to conclude that $C^2 = 1/2\pi$, giving finally:

$$(23) \qquad \left(\frac{\partial H(x,s)}{\partial s}\right)_{s=0} = \log \frac{\Gamma(x)}{\sqrt{2\pi}},$$

a formula discovered by Lerch in 1894 (see [7c] of the bibliography of this chapter). Finally, in the same way as we obtained (22), we get

$$S_1\left(x, 0, \frac{s+1}{2}\right) = H(x,s) - H(1-x, s),$$

from which we can now derive at once the value of $\partial S_1(x,0,s)/\partial s$ for $s = \tfrac{1}{2}$.

For $x = 1$, (23) gives the value of $\zeta'(0)$. More generally, one can obviously deduce from it the value of $L'(0)$ if $L(s)$ is any one of Dirichlet's L-functions; we shall make use of this in Chap. IX. Also, combining the above result on the value of $\partial S_1(x,0,s)/\partial s$ at $s = \tfrac{1}{2}$ with the functional equation (15), one gets at once Kummer's famous formula (obtained by him in 1847) for $\log \Gamma(x)$ in the interval $0 < x < 1$. Conversely, one can also, but somewhat more laboriously, deduce (23) from Kummer's formula and the results in § 7.

§ 10. The theory of distributions allows a re-interpretation of some of the above results; this will now be sketched briefly.

As is well-known, there is, on **R**, on **C** and more generally on every local field, one and (up to a constant factor) only one distribution which, under the action of the multiplicative group of that field, gets multiplied by a given quasicharacter ω of the latter group; such a distribution (which, suitably normalized, is sometimes known as "the Tate distribution") will be said to belong to ω.

---

[8] See e.g. E. Artin, *Einführung in die Theorie der Gammafunktion*, Hamb. Math. Einzelschr. no. 11, 1931.

Here we are dealing with **R**; the quasicharacters of $\mathbf{R}^\times$ are the functions of the form

$$x \mapsto \omega(x) = (\operatorname{sgn} x)^a |x|^z$$

with $a = 0$ or 1 and $z \in \mathbf{C}$. Assume first $\omega(x) = x^{-n}$, with an integer $n \geq 0$; then a distribution belonging to ω is $(d^n/dx^n)_{x=0}$, with its support at $x = 0$. Otherwise such a distribution is given as a linear function on the "Schwartz space" by the formula

(24) $$D_\omega(\Phi) = \operatorname{Pf} \int_{-\infty}^{+\infty} \Phi(x)\omega(x)|x|^{-1} dx;$$

here Φ is a "Schwartz function" (i.e. indefinitely differentiable and "rapidly decreasing"); the symbol Pf ("finite part") means the integral itself, whenever it is absolutely convergent; in general it is defined as the limit of

(25) $$\left(\int_{-\infty}^{-\varepsilon} + \int_{\varepsilon}^{+\infty} \right) \Phi(x)\omega(x)|x|^{-1} dx - \varepsilon^{a+z} P(\varepsilon^2) \qquad (\varepsilon > 0)$$

for $\varepsilon \to 0$ when P is a polynomial, so chosen that this has a finite limit; there is always such a polynomial, because Φ is indefinitely differentiable near $x = 0$.

The right-hand side of (24) is well defined provided: (a) Φ is everywhere locally integrable; (b) Φ is m times continuously differentiable near $x=0$, with $m > -\operatorname{Re}(z)$; (c) at $\pm \infty$, Φ is $O(|x|^\lambda)$ with $\lambda < -\operatorname{Re}(z)$. Assume now $\operatorname{Re}(z) < 0$; let φ be a function on $T = \mathbf{R}/\mathbf{Z}$, bounded, integrable, and indefinitely differentiable near 0; write Φ for the function on **R** given by

$$\Phi(x) = \varphi(x \bmod 1) \mathbf{e}(-xy),$$

and put $\Delta(\varphi) = D_\omega(\Phi)$. Then it is obvious that Δ is a distribution on T. For every continuous function φ whose support is disjoint from 0, we have

$$\Delta(\varphi) = \int_0^1 \varphi(x \bmod 1) \mathbf{e}(-xy) S_a\left(x, y, \frac{a+1-z}{2}\right) dx,$$

where the integrand is a periodic function of x of period 1; as usual, one expresses this by saying that, outside 0, Δ coincides with the function

(26) $$\mathbf{e}(-xy) S_a\left(x, y, \frac{a+1-z}{2}\right).$$

On the other hand, the coefficients of the Fourier series for Δ are given by

$$d_v = \Delta[\mathbf{e}(-vx)] = D_\omega[\mathbf{e}(-x(y+v))].$$

It is easily seen that $D_\omega(1)=0$, so that $d_\nu=0$ if $y+\nu=0$. Otherwise, since D_ω belongs to the quasicharacter ω, we have

$$d_\nu = D_\omega[\mathbf{e}(-x)]\,\omega(y+\nu)^{-1};$$

therefore, if we put $A_\omega = D_\omega[\mathbf{e}(-x)]$, the Fourier series for Δ is $A_\omega S_a\left(y, -x, \dfrac{a+z}{2}\right)$. As to A_ω, the easiest way of calculating it is to apply the definition (25) and to shift the integration to the imaginary axis in the complex x-plane; details may be left as an exercise to the reader. One finds:

$$A_\omega = (2\pi)^{-z}\,\Gamma(z)\,[\mathbf{e}(-z/4)+(-1)^a\mathbf{e}(z/4)].$$

Using well-known formulas for Γ, and putting $G(s)=\pi^{-s/2}\,\Gamma(s/2)$, we can also write this as

$$A_\omega = i^{-a}\pi^{\frac{1}{2}-z}\,\Gamma\!\left(\frac{a+z}{2}\right)\Gamma\!\left(\frac{a+1-z}{2}\right)^{-1} = i^{-a}\,G(a+z)\,G(a+1-z)^{-1}.$$

This suggests attaching to ω, not the distributions D_ω, Δ, but rather $G(a+z)^{-1}D_\omega$, $G(a+z)^{-1}\Delta$; in a more systematic theory, it would indeed be advisable to do so. In view of (26), the value we have just found for A_ω gives a formula which is formally identical with (15) of § 7, but with a different interpretation.

We will now write more explicitly Δ_ω instead of Δ for the distribution defined above for $\mathrm{Re}(z)<0$ and for ω not of the form $\omega(x)=x^{-n}$. Using the Fourier series for Δ_ω obtained above, it is now easily seen that, if φ is any indefinitely differentiable function on T, $G(a+z)^{-1}\Delta_\omega(\varphi)$, regarded as a function of z, can be continued analytically in the whole z-plane. This can be used in order to define $G(a+z)^{-1}\Delta_\omega$ as a distribution for all quasicharacters ω; its Fourier series is thus:

(27) $\qquad i^{-a}\,G(a+1-z)^{-1}\,S_a\!\left(y, -x, \dfrac{a+z}{2}\right).$

One of the most useful properties of distributions on T is that their Fourier series can always be differentiated term by term; this can of course be applied to the distributions introduced above; here we will merely illustrate it by a well-known example. For $\omega(x)=x$, $y=0$, the series (27) becomes

(28) $\qquad \sum{}' \dfrac{\mathbf{e}(\nu x)}{i\nu}.$

Differentiating it, we get the Fourier series which corresponds to the distribution $2\pi(\delta_0-1)$, where δ_0 is the mass 1 at 0 (the "Dirac distribution"). Therefore (28) is a function with a discontinuity at 0 and the constant derivative -2π outside 0; as it is an odd function of x, it must be $\pi(1-2\langle x\rangle)$. One should compare this with (7) of §4.

§ 11. Now we turn to series (4) for imaginary values of x. We will write:

$$(29) \qquad S_a(\zeta,y,s) = \sum_\mu (\bar{\zeta}+\mu)^a |\zeta+\mu|^{-2s} \mathbf{e}(-\mu y)$$

where a is an integer ≥ 0, $\zeta = \xi + i\eta$, ξ, η and y are real, and η is not 0. Since $\eta \neq 0$, it would not be enough here to restrict oneself to the cases $a=0$, $a=1$, as was done in § 7; on the other hand, the same assumption makes it unnecessary to use $\sum^*$. We begin by assuming $\operatorname{Re}(s) > \dfrac{a+1}{2}$; then (29) is absolutely convergent, and the function

$$S_a(\zeta,y,s)\mathbf{e}(-\xi y)$$

is periodic of period 1 in ξ, so that it can be expanded into a Fourier series. As (29) is introduced here merely because of its use in the study of Kronecker's double series in Chap. VIII, our concern will be only with that Fourier series. In a more systematic exposition, one would have to consider also, for instance, the relation between $S_a(\zeta,y,s)$ for imaginary ζ and $S_a(x,y,s)$ for real x, i.e. the behavior of $S_a(\zeta,y,s)$ when η tends to 0; as to this, one may consult Lipschitz' papers (cf. the bibliography of this chapter); it will be left aside here.

Whenever z is not real and ≤ 0, and t is complex, we will always take for z^t the value $e^{t\log z}$ with the "principal value" for $\log z$ (i.e. with $\operatorname{Im}(\log z)$ in the interval $]-\pi,\pi[$); thus we have $|z|^{2s} = z^s \bar{z}^s$. As in Chap. VI, we introduce the differential operators $\partial/\partial z$, $\partial/\partial \bar{z}$ for real-analytic (or real-differentiable) functions of $z \in \mathbf{C}$, with their usual meaning.

Let F be a function of $\zeta = \xi + i\eta$, defined in a strip $a < \eta < b$ and such that $F(\zeta)\mathbf{e}(-\xi y)$ is periodic of period 1 in ξ; expanding this into its Fourier series, we get for F a series:

$$F(\zeta) = \sum_v f_v(\eta)\mathbf{e}[(y+v)\xi] = \sum_v \int_0^1 F(x+i\eta)\mathbf{e}[(y+v)(\xi-x)]\,dx,$$

which, by "abuse of language", may be called the Fourier series for F. Call $F_v(\zeta)$ its v-th term. The operator $F \mapsto F_v$ commutes with translations in the ζ-plane, hence also with $\partial/\partial \xi$, $\partial/\partial \eta$ (or, what amounts to the same, with $\partial/\partial \zeta$, $\partial/\partial \bar{\zeta}$); it commutes also with $F \mapsto \varphi F$ if φ is any function of η. Consequently, if D is a linear differential operator in the ζ-plane whose coefficients are functions of η alone, $D(F)=0$ implies $D(F_v)=0$ for all v. Applying this to $F = S_a(\zeta,y,s)$ and to the operator

$$D = 2i\eta \frac{\partial^2}{\partial \zeta \partial \bar{\zeta}} + (a-s)\frac{\partial}{\partial \zeta} + s\frac{\partial}{\partial \bar{\zeta}},$$

we find that in this case the coefficient f_v of the Fourier series must satisfy the differential equation

$$(30) \qquad \eta \frac{d^2 f_v}{d\eta^2} + (2s-a)\frac{df_v}{d\eta} + [2a\pi(y+v) - 4\pi^2(y+v)^2 \eta] f_v = 0.$$

Moreover $\left(\text{always under the assumption } \operatorname{Re}(s) > \frac{a+1}{2}\right)$ it is obvious that f_v tends to 0 for $|\eta| \to \infty$. For instance, if $y+v=0$, f_v must be of the form $C|\eta|^{1+a-2s}$, where C is constant for $\eta > 0$ and also for $\eta < 0$; this is obvious anyway by homogeneity. For $y+v \neq 0$, it can still be shown, by using some elementary properties of Bessel functions, that (30) determines f_v uniquely, up to a similar factor C, and that it can be expressed in terms of the function $K_{s-a-\frac{1}{2}}$ and its derivatives. We proceed now to do the same, with the explicit determination of the factor C, by using Fourier's formulas for the coefficients f_v.

§ 12. In doing this, it is convenient to begin with the case $a=0$; for $a>0$, $\operatorname{Re}(s) > a + \frac{1}{2}$, one can then use the formula

$$(31) \qquad S_a(\zeta, y, s) = \frac{(-1)^a}{(s-1)(s-2)\ldots(s-a)} \frac{\partial^a}{\partial \zeta^a} S_0(\zeta, y, s-a)$$

to derive the Fourier series for $S_a(\zeta, y, s)$ for that case; finally, one uses analytic continuation in the s-plane to establish the corresponding formula for $\frac{a+1}{2} < \operatorname{Re}(s) \leq a + \frac{1}{2}$, and also to continue $S_a(\zeta, y, s)$ analytically even for $\operatorname{Re}(s) \leq \frac{a+1}{2}$.

Now we have to calculate f_v for $S_0(\zeta, y, s)$ by applying Fourier's formulas; this, of course, amounts to the same as applying Poisson summation to the series (29) for $a=0$. We get

$$(32) \qquad S_0(\zeta, y, s) = \sum_v \varphi(\eta, y+v) e[(y+v)\xi]$$

with φ defined by

$$(33) \qquad \varphi(\eta, y) = \int_{-\infty}^{+\infty} |\zeta|^{-2s} e(-y\xi) d\xi.$$

This is essentially a classical integral, with a long history behind it[9]. Multiplying with $\Gamma(s)$, one can apply to it Dirichlet's transformation (cf. § 2):

---

[9] Cf. G. N. Watson, *Theory of Bessel Functions*, § 6.16 (pp. 172—173), where one will find some bibliography.

$$\text{(34)} \quad \Gamma(s)\varphi(\eta,y) = \iint \exp[-t(\xi^2+\eta^2) - 2\pi i y\xi] t^{s-1} dt\, d\xi$$
$$= \sqrt{\pi} \int_0^{+\infty} \exp\left(-t\eta^2 - \frac{\pi^2 y^2}{t}\right) t^{s-\frac{3}{2}} dt.$$

For $y=0$, we get:

$$\text{(35)} \quad \varphi(\eta,0) = |\eta|^{1-2s} \sqrt{\pi} \frac{\Gamma(s-\frac{1}{2})}{\Gamma(s)}$$

For $y \neq 0$, the integral is essentially one which can be used to define the so-called K-function (cf. loc. cit.[9], § 6.22, p. 183, and the bibliography given there); it goes back to Poisson. One puts, for any complex z, and for $\operatorname{Re}(Y) > 0$:

$$\text{(36)} \quad K_z(2Y) = \frac{1}{2} \int_0^{+\infty} \exp\left[-Y\left(t+\frac{1}{t}\right)\right] t^{z-1} dt;$$

for each Y, this is an even entire function of z. Now (34) gives

$$\text{(37)} \quad \Gamma(s)\varphi(\eta,y) = 2\sqrt{\pi} |\pi y/\eta|^z K_z(2|\pi y\eta|) = 2\sqrt{\pi} |\pi y|^{2z} Y^{-z} K_z(2Y)$$
$$(y \neq 0,\ z = s - \tfrac{1}{2},\ Y = |\pi y\eta|).$$

One can now write down the Fourier series for $S_0(\zeta, y, s)$ and then, after replacing s by $s-a$ in that series, use (31) to write down the series for $S_a(\zeta, y, s)$ for $\operatorname{Re}(s) > a + \tfrac{1}{2}$; the calculation is a routine one, and it will be enough to state the result. For $\varepsilon = \pm 1$, $Y > 0$, define

$$\text{(38)} \quad \Phi_\varepsilon(Y, a, z) = e^{2\varepsilon Y} \frac{d^a}{dY^a}\left[e^{-2\varepsilon Y} Y^{-z} K_z(2Y)\right].$$

The Fourier series for S_a is:

$$\text{(39)} \quad S_a(\zeta, y, s) = 2(\delta i)^{-a} \Gamma(s)^{-1} \sum_v C_v \mathbf{e}[(y+v)\xi]$$

with $\delta = \operatorname{sgn} \eta$, and coefficients C_v given as follows. If $y + v = 0$, then we have

$$\text{(40)} \quad C_v = \pi \frac{\Gamma(2s-a-1)}{\Gamma(s-a)} |2\eta|^{a+1-2s};$$

otherwise we have

$$\text{(41)} \quad C_v = \sqrt{\pi}(-2)^{-a} |\pi(y+v)|^{2s-a-1} \Phi_\varepsilon(|\pi(y+v)\eta|, a, s-a-\tfrac{1}{2})$$

with $\varepsilon = \operatorname{sgn}[(y+v)\eta]$.

For $Y \to +\infty$, the function K_z has a well-known asymptotic expansion (cf. Watson, loc. cit., § 7.23, p. 202); but all we need here is the easily established fact that, for all z, $K_z(2Y)$ and all its derivatives are $O(e^{-\lambda Y})$ for some $\lambda > 0$ (actually, any $\lambda < 2$); consequently, the same holds for Φ_g, and the series in (39) converges (as fast as a geometric series) for all values of the parameters. Using the principle of analytic continuation, we conclude that (39) remains valid whenever the left-hand side is defined, i.e. for $\mathrm{Re}(s) > \frac{a+1}{2}$; moreover, we can use (39) to define $S_a(\zeta, y, s)$ for all values of a, ζ, y, s, except when y is an integer and the constant coefficient in (39), as given by (40), is ∞.

§ 13. Consider a positive-definite binary quadratic form $F(X, Y)$ with real coefficients, and write it as

$$F(X,Y) = AX^2 + BXY + CY^2 = A(X + \zeta Y)(X + \overline{\zeta} Y).$$

Then, instead of the series $S_a(\zeta, y, s)$, one could have considered the series

$$A^{-s} S_a(\zeta, y, s) = \sum_\mu (\mu + \overline{\zeta})^a F(\mu, 1)^{-s} \mathbf{e}(-\mu y);$$

except for the factor A^{-s}, one would of course get the same results as above.

At various times, in connection with his double series, Kronecker seems to have attached some importance to the fact that many results concerning positive-definite forms $F(X, Y)$ remain valid for complex-valued forms with positive-definite real part. This will now be explained briefly. Write the form in question as

$$F(X,Y) = AX^2 + BXY + CY^2 = A(X + \zeta Y)(X + \zeta' Y);$$

we assume that $\mathrm{Re}[F(X,Y)] > 0$ for all real X, Y, so that ζ, ζ' cannot be real. For all real t, $F(t, 1)$ remains in the half-plane $\mathrm{Re}(z) > 0$. If $\mathrm{Im}(\zeta)$ and $\mathrm{Im}(\zeta')$ were both positive, then the arguments of $t + \zeta$, $t + \zeta'$ would decrease from π to 0, and that of $A^{-1} F(t, 1)$ from 2π to 0, as t varies from $-\infty$ to $+\infty$, against our assumption[10]; for a similar reason, $\mathrm{Im}(\zeta)$ and $\mathrm{Im}(\zeta')$ cannot both be negative.

Conversely, if $\mathrm{Im}(\zeta)$ and $\mathrm{Im}(\zeta')$ are of opposite signs, the argument of $(t + \zeta')(t + \overline{\zeta})^{-1}$, which is the same as that of $A^{-1} F(t, 1)$, varies from 0 back to 0 when t varies from $-\infty$ to $+\infty$; as that point moves on a circle in the complex plane, it follows that it remains inside a fixed half-plane, i.e. that one can choose A so that $\mathrm{Re}[F(t, 1)] > 0$ for all real t. Putting now $\eta = \frac{1}{2i}(\zeta - \zeta')$, $\delta = \mathrm{sgn}\, \mathrm{Re}(\eta)$, one sees easily that $A = 1/\delta \eta$ is such a choice. We will also write $\xi = \frac{1}{2}(\zeta + \zeta')$; for

---

[10] Here and at other places, in this and the next chapters, we have borrowed freely from C. L. Siegel's *Lectures on advanced analytic number-theory* (T.I.F.R., Bombay 1961).

given values of ξ, η, the above conditions for ζ, ζ' are satisfied if and only if $|\text{Im}(\xi)| < |\text{Re}(\eta)|$; if η is given, this condition determines a strip in the complex ξ-plane. We will also write $\eta' = \delta \eta$, so that $\text{Re}(\eta') > 0$ and $A = \eta'^{-1}$.

With these notations, we will consider the extension of the results of § 12 to the series

$$S'_a = \eta'^{-s} \sum_\mu (\mu + \zeta')^a F(\mu, 1)^{-s} \mathbf{e}(-\mu y),$$

where y is real; for $\zeta' = \overline{\zeta}$, this becomes identical with (29). For a given η, $S'_a \mathbf{e}(-\xi y)$ is a holomorphic function of ξ, periodic of period 1, in the strip of the ξ-plane mentioned above. Just as in §§ 11—12, writing down the Fourier series for that function amounts to applying Poisson summation to S'_a; moreover, S'_a can be expressed in terms of S'_0 by a formula which is formally identical with (31).

As to S'_0, the Fourier formulas give

$$S'_0 = \eta'^{-s} \sum_v \psi(\eta', y+v) \mathbf{e}[(y+v)\xi],$$

with ψ defined by the integral

$$\psi(\eta', y) = \int_{-\infty}^{+\infty} \left(\eta' + \frac{t^2}{\eta'}\right)^{-s} \mathbf{e}(-ty) dt$$

taken along the real t-axis. Applying to this the Dirichlet transformation, one gets, for $y \neq 0$:

$$\Gamma(s)\psi(\eta', y) = 2\sqrt{\pi\eta'} |\pi y|^z K_z(2Y) \qquad (z = s - \tfrac{1}{2},\ Y = \pi |y|\eta'),$$

and, for $y = 0$:

$$\Gamma(s)\psi(\eta', 0) = \eta'^{1-s} \sqrt{\pi} \Gamma(s - \tfrac{1}{2}).$$

The calculation then proceeds exactly as in § 12, except that $|\eta|$ has now to be replaced by η', and that one has to take $\varepsilon = \delta \,\text{sgn}(y+v)$; then (39) remains valid as before. Also the estimates concerning the convergence of that series remains as before, in view of the obvious inequality

$$|K_z(2Y)| \leq K_{\text{Re}(z)}[2\,\text{Re}(Y)].$$

Bibliography

[1] Malmstén, C.J.: *De integralibus quibusdam definitis, seriebusque infinitis*, Crelles J. **38**, 1—39 (1849).
[2] Lipschitz, R.: *Untersuchung einer aus vier Elementen gebildeten Reihe*, Crelles J. **54**, 313—328 (1857)

[3] Hurwitz, A.: *Einige Eigenschaften der Dirichlet'schen Funktionen* $F(s) = \sum \left(\dfrac{D}{n}\right) \dfrac{1}{n^s}$, *die bei der Bestimmung der Klassenzahlen binärer quadratischer Formen auftreten*, Zeitschr. für Math. u. Phys. **27**, 86—101 (1882) (= Math. Werke Bd. I, 72—88).

[4] Lerch, M.: *Note sur la fonction* $\mathfrak{R}(w, x, s) = \sum\limits_{0}^{\infty} \dfrac{e^{2k\pi i x}}{(w+k)^s}$, Acta Math. **11**, 19—24 (1887—88).

[5] Lipschitz, R.: *Untersuchung der Eigenschaften einer Gattung von unendlichen Reihen*, Crelles J. **105**, 127—156 (1889).

[6] Lerch, M.: *Sur certains développements en séries trigonométriques*, Ann. Toulouse (I) **3**, 1—11 (1889).

[7] Lerch, M.: (a) *Grundzüge der Theorie der Malmsténschen Reihen*, Rozpravy česke akad. I, no. 27 (1892); (b) *Studien auf dem Gebiete der Malmstén'schen Reihen*, ibid. II, nos. 4, 23 (1893); (c) *Weitere Studien...*, ibid. III, no. 28 (1894) [for these papers, written in czech, cf. the author's summaries, Jahrbuch über die Fortschr. d. Math. 1892, pp. 446—452; 1893—94, pp. 790—793 and 484—486].

[8] Lerch, M.: *Über den Kronecker'schen Beweis der sogenannten Kronecker'schen Grenzformel*, Arch. d. Math. u. Phys. (III) **6**, 85—94 (1904).

Chapter VIII

Kronecker's Double Series

§ 1. As in earlier chapters, we write W for a lattice in the complex plane, and u, v for two generators of W, so that W consists of the points $w = \mu u + v v$, where μ, v are integers. By χ we will denote a character of the additive group W; occasionally we shall write $\chi(\mu, v)$ for $\chi(\mu u + v v)$.

The double series with which Kronecker was dealing in his later years were all of the form

(1) $$\sum \chi(w)(\bar{x}+\bar{w})^a |x+w|^{-2s},$$

and he introduced most of the basic tools which can be applied to the theory of such series. It is only Lerch, however, who, under the influence of Kronecker's work, introduced series of that general type. Kronecker himself, until 1889, considered only the case $a=0$, $x=0$ (i.e. the series (1) of Chap. VII, § 1), chiefly with a view to studying the series at $s=1$. In 1890 and 1891 (particularly in the latter year, in part under the influence of his rediscovery of Eisenstein's work) he began to give special emphasis to the case $a=s=1$, i.e. to the series

(2) $$\sum_{w \in W} \frac{\chi(w)}{x+w}$$

which he appears to have regarded at that time as the cornerstone of the whole theory of elliptic functions (*Werke* V, pp. 103—104). We begin with an exposition of his main results on that series.

In dealing with it, Kronecker tried various methods of summation, including the method

$$\lim_{N \to +\infty} \left(\sum_{\mu=-N}^{N} \sum_{v=-N}^{N} \right)$$

and even a variant of "Kronecker summation" based on the strange series $\sum \chi(w)(x+w)^{-s}$ (*Werke* V, pp. 104—127). Eventually he found Eisenstein summation to be the most appropriate procedure. That is the method we shall follow here.

§ 2. As in Chapters III and IV, we put

$$\zeta = x/u, \quad \tau = \delta v/u, \quad z = \mathbf{e}(\zeta), \quad q = \mathbf{e}(\tau).$$

Moreover, we put:

$$\chi(u) = \mathbf{e}(-\delta\beta_0), \quad \chi(v) = \mathbf{e}(\delta\alpha_0), \quad x_0 = \alpha_0 u + \beta_0 v,$$
$$\zeta_0 = x_0/u, \quad z_0 = \mathbf{e}(\zeta_0).$$

With these notations, the series (2), suitably summed, is what Kronecker denotes by $\mathrm{Ser}(x_0, x, u, v)$. We assume that x is not in W, and also, for the moment, that $\chi(u) \neq 1$; then we can choose β_0 so that $0 < \delta\beta_0 < 1$. Since $|z_0| = |q|^{\delta\beta_0}$, we have then $1 > |z_0| > |q|$.

Formula (6) of Chap. VII, § 4, gives now

$$\sum_{\mu} \frac{\chi(\mu u + vv)}{x + \mu u + vv} = \frac{2\pi i}{u} \mathbf{e}\left(\frac{\delta\beta_0 x}{u}\right) \frac{z_0^{\delta v}}{q^{\delta v} z - 1};$$

the series in the left-hand side is convergent (in the "ordinary" sense). As $1 > |z_0| > |q|$, $\sum_{v}$, applied to the right-hand side, is absolutely convergent. We get

(3) $$\sum_{e} \frac{\chi(w)}{x + w} = \frac{2\pi i}{u} \mathbf{e}\left(\frac{\delta\beta_0 x}{u}\right) F(q, z, z_0)$$

where F is the function defined, for $1 > |w| > |q|$, by the series

(4) $$F(q, z, w) = \sum_{v} \frac{w^v}{q^v z - 1};$$

for $1 > |z| > |q|$, this may also be written more symmetrically:

$$F(q, z, w) = 1 - \frac{1}{1-z} - \frac{1}{1-w} + \sum_{m=1}^{+\infty} \sum_{n=1}^{+\infty} q^{mn}(z^{-m}w^{-n} - z^m w^n).$$

§ 3. Now, following Kronecker fairly closely (*Werke* IV, pp. 309—318), we will show that F can be expressed, in terms of the infinite products $X_q(z)$, $P(q)$ defined in Chap. IV, § 3, by the formula:

(5) $$F(q, z, w) = \frac{P(q)^2 X_q(zw)}{X_q(z) X_q(w)}.$$

For a given q, write $\Phi(z,w)$ for the right-hand side. It is obvious that $w \mapsto \Phi(z,w)$ is a meromorphic function of w for $w \neq 0$, with poles at the zeros of X_q, i.e. at the points $w = q^v$; consequently, for $1 > |w| > |q|$, it can be expanded into a Laurent series

$$\Phi(z,w) = \sum_v f_v(z) w^v.$$

The formula (19) of Chap. IV, § 5, gives:

(6) $$\Phi(z,w) = w \Phi(qz,w) = z \Phi(z,qw).$$

The first one of these relations gives, for all v:

$$f_v(z) = f_0(q^v z).$$

The second one shows that, for $|q|^{-1} > |w| > 1$, we have

$$\Phi(z,w) = \sum_v z f_v(z) q^v w^v.$$

Let γ, γ' be the circles $|w| = |q|^{\frac{1}{2}}$, $|w| = |q|^{-\frac{1}{2}}$ in the w-plane, taken in the positive sense. The integrals of $(2\pi i w)^{-1} \Phi(z,w) dw$, taken along γ and γ', have the values $f_0(z), z f_0(z)$, respectively. By Cauchy's theorem, the difference of the two integrals is the residue of $w^{-1} \Phi$ at $w = 1$; as $P(q)^2$ is the derivative of $X_q(w)$ at $w = 1$, this residue is 1. This gives $(z-1) f_0(z) = 1$ and completes the proof for (5).

§ 4. Accordingly, we define now $F(q,z,w)$ by (5) for all values of z, w. Replace x_0 by $x_0 + w_0$ with $w_0 \in W$; this changes z_0 into $q^v z_0$ with an integer v; in view of (6), the right-hand side of (3) is then unchanged; in particular, this right-hand side is periodic of period 1 with respect to β_0. Consequently, (3) remains valid provided $\chi(u) \neq 1$. If $\chi \neq 1$ but $\chi(u) = 1$, then the left-hand side of (3) ceases to be convergent even in the sense of Eisenstein summation; (3) would remain valid, with $F(q,z,w)$ still defined by (5), if one replaced $\sum_e$ by a suitably modified summation process, as one could easily verify by making use of the results of Chap. II.

Making use of (15), Chap. IV, § 3, one can rewrite (3) and (5) as

(7) $$\sum_e \frac{\chi(w)}{x+w} = \mathbf{e}\left(\frac{\delta \beta_0 x}{u}\right) \frac{\varphi(x+x_0)}{\varphi(x)\varphi(x_0)},$$

this being valid for $\chi(u) \neq 1$. Applying (17) of Chap. IV, § 4, one verifies easily that the right-hand side does not depend upon the choice of the generators u, v for the lattice W.

§ 5. This last result can also be obtained as follows. Put

$$\lambda(w,s) = \chi(w)(\bar{x}+\bar{w})|x+w|^{-2s}$$

and consider the series $\sum \lambda(w,s)$; for $s=1$, it is formally the same as the series in the left-hand side of (7). Assume that x is not in W and that $\chi(u) \neq 1$. Making use of Chap. VII, § 4 and § 11, we can write:

(8) $$\sum_\mu \lambda(\mu u + vv, s) = \chi(vv)\bar{u}|u|^{-2s} S_1(\zeta + \delta v\tau, \delta \beta_0, s);$$

here the left-hand side converges (in the "ordinary" sense) for $\mathrm{Re}(s) > \tfrac{1}{2}$, as we proved in Chap. VII, § 4; the right-hand side is defined by (29) of Chap. VII, § 11, for $\mathrm{Re}(s) > 1$, and by (39) and (41) of Chap. VII, § 12, for arbitrary s; (40) of Chap. VII, § 12, is not needed here, since x is not in W and therefore $\zeta + \delta v\tau$ cannot be in $\mathbf{Z}$. For $\mathrm{Re}(s) > 1$, (8) is an identity; therefore it remains valid, by analytic continuation, for $\mathrm{Re}(s) > \tfrac{1}{2}$.

Apply now $\sum_v$ to (8). Replacing S_1 in the right-hand side by the series (39) of Chap. VII, § 12, we obtain a double series which is easily seen to be absolutely convergent for all s; this requires no more than the crude estimate given for the Bessel function K_z (and, as a consequence, for Φ_ε) at the end of Chap. VII, § 12. Thus we get:

$$\sum_e \frac{\chi(w)}{x+w} = \sum_v \left(\sum_\mu \lambda(w,1) \right) = \lim_{s=1} \sum_v \left(\sum_\mu \lambda(w,s) \right),$$

provided x is not in W, and $\chi(u) \neq 1$; in the right-hand side, the limit is taken for $s > 1$, $s \to 1$. As the double series in the right-hand side is absolutely convergent for $s > \tfrac{3}{2}$ and therefore independent of the choice of the generators u, v for W, the same remains true under analytic continuation along the real axis in the s-plane; therefore it is true of the left-hand side.

§ 6. Now we turn to Kronecker's series

$$G(s,\chi) = \sum{}' \chi(w)|w|^{-2s},$$

where $\sum'$, as usual, denotes the sum taken over all $w \in W$ except 0. As before, write $w = \mu u + vv$, $|w|^2 = F(\mu, v)$, $\chi(w) = \chi(\mu, v)$; then we can write:

(9) $$G(s,\chi) = \sum{}' \chi(\mu,v) F(\mu,v)^{-s};$$

here F is a positive-definite quadratic form. Conversely, any such form can be written as $|\mu u + vv|^2$. The lattice W admits a complex multiplier (in the sense of Chap. V, § 6) if and only if F can be written as $F = cF_1$, where F_1 is a form with coefficients in $\mathbf{Z}$.

Series such as (9), for the case where $\chi=1$ and F has its coefficients in **Z**, appear for the first time in Dirichlet's work on the class-number of binary quadratic forms; he proved that such a series has at $s=1$ a simple pole with the residue $2\pi D^{-\frac{1}{2}}$ if $-D$ is the discriminant of F. The proof remains valid even when the coefficients of F are not integers.

Kronecker's main results concern (a) the value of $G(1,\chi)$ for $\chi\neq 1$, and (b) the constant term in the expansion of $G(s,1)$ at $s=1$; the latter is known as his "limit-formula" ("*Kroneckersche Grenzformel*") or sometimes as his "first" limit-formula, while the former is called his "second" limit-formula. He deduced (b) from (a), at first (*Werke* IV, pp. 376—379) for forms with integral coefficients, and later (*Werke* IV, pp. 482—495) in the general case[11]. As he remarked on various occasions, one can also generalize such results to the case when F is a binary quadratic form with positive-definite real part (cf. above, Chap. VII, § 13); this presents no difficulty if one makes use of the results described in our Chap. VII, § 13, and it will not be considered here.

§ 7. Later writers (M. Lerch, H. Weber, etc.) found that it is even easier to treat $G(s,1)$ directly. Their proofs seem to differ from one another only superficially; here we will follow the presentation in Chowla-Selberg, *Crelles J.* 227 (1967), pp. 86—110. With the notations of Chap. VII, § 4 and § 11, we have:

$$(10) \qquad G(s,1) = |u|^{-2s} \sum_{\mu,\nu} |\mu + \nu\tau|^{-2s} = |u|^{-2s} \sum_{\nu} S_0(\nu\tau, 0, s),$$

where the series are absolutely convergent for $\text{Re}(s) > 1$.

Here $S_0(0,0,s)$ is no other than $2\zeta(2s)$; formula (15) of Chap. VII, § 7, gives its well-known functional equation; in the same §, (14) shows that it has the unique pole $s=\frac{1}{2}$, with the residue 1; then (15) gives $2\zeta(0) = -1$, and (23) of Chap. VII, § 9, gives $2\zeta'(0) = -\log(2\pi)$; finally, using (15) again, together with the known facts about $\Gamma(s)$, one finds that the constant term in the expansion of $\zeta(2s)$ at $s=\frac{1}{2}$ is Euler's constant $\gamma = -\Gamma'(1)$.

Now consider the terms in the right-hand side of (10) for $\nu \neq 0$. Put $\tau = \xi + i\omega$; according to the definition of τ, we have $\omega > 0$. For S_0, we use the formulas (32), (35), (37) of Chap. VII, § 12; this gives:

$$(11) \qquad \begin{aligned} S_0(\nu\tau, 0, s) &= \sqrt{\pi}\, \Gamma(s-\tfrac{1}{2})\Gamma(s)^{-1} |\omega\nu|^{1-2s} \\ &\quad + 2\pi^s \Gamma(s)^{-1} \sum_{\rho}' |\rho/\omega\nu|^z K_z(2\pi\omega|\nu\rho|) \mathbf{e}(\nu\rho\xi), \end{aligned}$$

with $z = s - \frac{1}{2}$. Substituting this in (10), we get

---

[11] A general proof of the "first limit-formula" had been obtained by H. Weber (inspired by Kronecker's paper of 1863, *Werke* IV, pp. 221—225) as early as 1881; cf. Dedekind, *Werke* II, p. 225; it is the proof published in *Math. Ann.* 33 (1889), pp. 392—395.

(12) $$|u|^{2s}G(s,1)=2\zeta(2s)+2\sqrt{\pi}\,\Gamma(s-\tfrac{1}{2})\Gamma(s)^{-1}\omega^{1-2s}\zeta(2s-1)$$
$$+2\sqrt{\pi}\,\Gamma(s)^{-1}(\pi/\omega)^{z}G_{1}(z)$$

where $G_1(z)$ is the double series

(13) $$G_{1}(z)=\sum_{\nu}{}'\sum_{\rho}{}'|\rho/\nu|^{z}K_{z}(2\pi\omega|\nu\rho|)\mathbf{e}(\nu\rho\xi).$$

Using the estimate for K_z at the end of Chap. VII, § 12, one sees that the double series for G_1 is absolutely convergent for all z and represents an entire function of z. It is an even function of z, as one sees by exchanging ν and ρ in (13).

§ 8. Taking into account the functional equation for ζ, and the known facts about $\Gamma(s)$, one can write (12) more symmetrically:

(14) $$|u|^{2s}G(s,1)=2\zeta(2s)+2\Gamma(1-s)\Gamma(s)^{-1}(\pi/\omega)^{2s-1}\zeta(2-2s)$$
$$+2\sqrt{\pi}\,\Gamma(s)^{-1}(\pi/\omega)^{z}G_{1}(z).$$

Multiplying with $\Gamma(s)(\pi/\omega)^{-z}$, we get an even function of z; this is the "functional equation" for $G(s,1)$; a more general result will be obtained in § 13. We also see that $G(s,1)$ has no pole in the s-plane except $s=1$, and that its residue there is $\pi/(\omega u\bar{u})$, which can also be written as $2\pi D^{-\frac{1}{2}}$, if $-D$ is the discriminant of the quadratic form $F(\mu,\nu)$; this had been Dirichlet's result. Formula (14) also gives $G(0,1)=-1$.

In order to get the next term in the expansion of $G(s,1)$, either at $s=0$ or at $s=1$, we have to calculate $G_1(\tfrac{1}{2})$. It is well-known that $K_{\frac{1}{2}}(2Y)=\tfrac{1}{2}(\pi/Y)^{\frac{1}{2}}e^{-2Y}$; it may be verified easily by making the change of variable $t^{\frac{1}{2}}-t^{-\frac{1}{2}}=\theta$ in the integral (36) of Chap. VII, § 12. This gives:

$$\sqrt{\omega}\,G_{1}(\tfrac{1}{2})=\sum_{\nu=1}^{+\infty}\sum_{\rho=1}^{+\infty}\frac{1}{\nu}(q^{\nu\rho}+\bar{q}^{\nu\rho})=-\log[P(q)P(\bar{q})];$$

here we have put $q=\mathbf{e}(\tau)$ as usual; $\bar{q}=\mathbf{e}(-\bar{\tau})$ is the complex-conjugate of q; $P(q)$ is the infinite product introduced in (14) of Chap. IV, § 3.

Using the above-mentioned facts about $\zeta(s)$, and the known properties of $\Gamma(s)$, we can now write the first two terms in the expansions of $G(s,1)$ at $s=1$ and at $s=0$. It will be convenient to express them in terms of the "discriminant" $\Delta=\Delta(W)$ defined by (36) of Chap. IV, § 11, and also to introduce again (as in Chap. VI, § 2) the notation:

(15) $$u\bar{v}-v\bar{u}=-\delta u\bar{u}(\tau-\bar{\tau})=-2i\delta\omega u\bar{u}=-2\pi i\delta A,$$

where the constant $A=\omega u\bar{u}/\pi$ depends only upon the lattice W and not upon the choice of the generators u,v.

We can now write the expansion of $G(s,1)$ at $s=1$ as

(16) $$A \cdot G(s,1) = \frac{1}{s-1} + 2\gamma - \log(A^2) - \frac{1}{12}\log(\Delta\bar{\Delta}) + \cdots,$$

where γ is Euler's constant, and the omitted terms are 0 at $s=1$. This is Kronecker's limit-formula.

At $s=0$, we have

(17) $$G(s,1) = -1 - \frac{s}{12}\log(\Delta\bar{\Delta}) + \cdots.$$

In view of the functional equation for $G(s,1)$, the formulas at $s=1$ and at $s=0$ are equivalent. As often happens with arithmetically defined Dirichlet series, the formula at $s=0$ is the simpler one.

§ 9. Now we consider $G(s,\chi)$ for $\chi \neq 1$, particularly at $s=1$. In expressing $G(1,\chi)$ in terms of known functions, one has the choice between Eisenstein summation and Kronecker summation; as will be shown presently, both lead to the same result. Actually, when Kronecker dealt with this series for the first time (*Werke* IV, pp. 347—351), he was using (unwittingly, no doubt) the essential idea in Eisenstein summation.

Formally, with the notations explained in § 2, we can write:

(18) $$G(s,\chi) = \sum_{\mu,\nu}' F(\mu,\nu)^{-s} \chi(\mu u + \nu v)$$
$$= \sum_{\mu,\nu}' |u|^{-2s} |\mu + \delta\nu\tau|^{-2s} \mathbf{e}(-\delta\beta_0\mu + \delta\alpha_0\nu) = \sum_\nu A_\nu(s)$$

where we have put:

(19) $$A_\nu(s) = |u|^{-2s} S_0(\nu\tau, \delta\beta_0, s) \mathbf{e}(\alpha_0 \nu)$$

with S_0 defined, as always, by (4) of Chap. VII, § 4. For $\mathrm{Re}(s) > 1$ everything here is absolutely convergent.

Take $s=1$. From (8) of Chap. VII, § 4, we get:

(20) $$A_0(1) = \frac{2\pi^2}{u\bar{u}}\left(\beta_0^2 - \delta\beta_0 + \frac{1}{6}\right) \quad (0 \leq \delta\beta_0 < 1).$$

In order to calculate $A_\nu(1)$ for $\nu \neq 0$, Kronecker starts from the identity (essentially the same as (1) of Chap. II, § 2, which had been Eisenstein's point of departure):

(21) $$F(\mu,\nu)^{-1} = (\mu u + \nu v)^{-1}(\mu\bar{u} + \nu\bar{v})^{-1}$$
$$= (2i\delta\omega u\bar{u})^{-1}\left(\frac{\nu^{-1}}{\mu + \delta\nu\bar\tau} - \frac{\nu^{-1}}{\mu + \delta\nu\tau}\right).$$

Actually, at this point (*Werke* IV, pp. 350—351), Kronecker deals with a quadratic form F with positive-definite real part (cf. Chap. VII, § 13); that case, more general than the one we are considering here, can be treated exactly in the same way.

Now, assuming at first $0 < \delta\beta_0 < 1$, we calculate $A_\nu(1)$ by applying (6) of Chap. VII, § 4, combined with (21). This gives:

$$(22) \quad A_\nu(1) = \sum_\mu F(\mu, \nu)^{-1} \chi(\mu u + \nu v) = A^{-1} [f_{\delta\nu}(\tau, \zeta_0) - f_{\delta\nu}(\bar{\tau}, \bar{\zeta}_0)]$$

with ζ_0 defined as in § 2, and the function f_ν defined, for all $\nu \neq 0$, by the formula

$$f_\nu(\tau, \zeta_0) = \nu^{-1} \frac{\mathbf{e}(\nu \zeta_0)}{1 - \mathbf{e}(\nu \tau)} = \frac{\nu^{-1} z_0^\nu}{1 - q^\nu}.$$

One should note that the series in (22) is absolutely convergent. As we have $\mathbf{e}(\bar{\zeta}_0) = \bar{z}_0^{-1}$, $\mathbf{e}(\bar{\tau}) = \bar{q}^{-1}$, we have:

$$f_{-\nu}(\bar{\tau}, \bar{\zeta}_0) = -\frac{\nu^{-1} \bar{z}_0^\nu}{1 - \bar{q}^\nu}.$$

Now the infinite product $X_q(z)$, defined in Chap. IV, § 3, can be written as

$$X_q(z) = -z^{-\frac{1}{2}} \prod_{n=0}^{+\infty} (1 - q^n z)(1 - q^{n+1} z^{-1}).$$

If $1 > |z| > |q|$, this gives:

$$\log(-z^{\frac{1}{2}} X_q(z)) = -\sum_{n=0}^{+\infty} \sum_{\nu=1}^{+\infty} \nu^{-1}(q^{n\nu} z^\nu + q^{(n+1)\nu} z^{-\nu}) = -\sum' f_\nu(\tau, \zeta),$$

where everything is absolutely convergent. If $|z| = 1$, $z \neq 1$, Abel's partial summation (cf. Chap. VII, § 4) shows that the right-hand side is still convergent in the sense of ordinary convergence; by continuity, the above formula is therefore still valid. In it, we replace q, z, first by q, z_0 and then by $\bar{q}, \bar{z}_0$; assuming now that $0 \leq \delta\beta_0 < 1$, we have $1 \geq |z_0| > |q|$; also, with notations as before, we have $\log|q| = -2\pi\omega$ and $\log|z_0| = -2\pi\delta\beta_0\omega$. Finally, we get, combining all the above formulas:

$$\sum'_e \chi(w)|w|^{-2} = A_0(1) + \sum'_\nu A_{\delta\nu}(1)$$

$$(23) \qquad = \frac{2\pi^2}{u\bar{u}} \left(\beta_0^2 + \frac{1}{6} \right) - A^{-1} \log [X_q(z_0) X_{\bar{q}}(\bar{z}_0)].$$

This has been established for $0 \leq \delta\beta_0 < 1$; but the left-hand side does not depend upon the choice of β_0 when χ is given; as to the right-hand side, it is easy to verify (by using (19) of Chap. IV, § 5) that it does not change if β_0 is replaced by β_0+1, and z_0 is replaced correspondingly by $q^\delta z_0$. Therefore (23) is valid without any restriction, provided $\chi \neq 1$.

§ 10. Now we will consider Kronecker summation. In (19), for $v \neq 0$, replace $S_0(v\tau, \delta\beta_0, s)$ by its value as given by (32) of Chap. VII, § 12; this is a series consisting of terms given by (37) of the same §, all of which contain the function K_z with $z = s - \frac{1}{2}$, and, if β_0 is an integer (i.e. if $\chi(u) = 1$), of one term given by (35) of that §. Collect all the terms containing K_z; they make up a double series, for which we will write $C(s)$. On the other hand, if $\chi(u) = 1$, collect all the terms given by (35); they make up the series

(24)
$$B(s) = \sum_v{}' |u|^{-2s} \mathbf{e}(\alpha_0 v) \sqrt{\pi} \Gamma(s - \tfrac{1}{2}) \Gamma(s)^{-1} |\omega v|^{1-2s}$$
$$= |u|^{-2s} \sqrt{\pi} \Gamma(s - \tfrac{1}{2}) \Gamma(s)^{-1} \omega^{1-2s} S_0(0, -\alpha_0, s - \tfrac{1}{2}).$$

If $\chi(u) \neq 1$, we put $B(s) = 0$.

Thus we have, for $\operatorname{Re}(s) > 1$:

(25)
$$\sum{}' \chi(w) |w|^{-2s} = A_0(s) + B(s) + C(s).$$

To evaluate $C(s)$, it is enough to use the crude estimate for K_z mentioned at the end of Chap. VII, § 12; this shows at once that the double series $C(s)$ is absolutely convergent for all s and represents an entire function of s. As to $A_0(s)$ and $B(s)$, we can apply to them (14) of Chap. VII, § 7, which shows that they are meromorphic in the whole s-plane. More precisely, $A_0(s)$ is entire unless $\chi(u) = 1$; if $\chi(u) = 1$, it has only one simple pole at $s = \frac{1}{2}$, with the residue $|u|^{-1}$. If $\chi(u) = 1$, $B(s)$ has a pole with the residue $-|u|^{-1}$ at $s = \frac{1}{2}$, and, if α_0 is an integer, i.e. if also $\chi(v) = 1$, it has an additional pole at $s = 1$, with the residue A^{-1}. If $\chi = 1$, this coincides with the result obtained in § 8; if not, it shows that the right-hand side of (25) is an entire function of s.

§ 11. For $\operatorname{Re}(s) > 1$, the left-hand side of (25) depends only upon the lattice W and the character χ, and not upon the choice of the generators u, v for W; therefore the same is true of the right-hand side for $\operatorname{Re}(s) > 1$, and, by analytic continuation, also for all s. Now consider the series $\sum A_v(s)$; if $\chi(u) \neq 1$, it is absolutely convergent for all s, since the series $C(s)$ is so. For $\chi(u) \neq 1$, its value for $s = 1$ is equal to the right-hand side of (23). In other words, "Kronecker summation" gives for the series defining $G(s, \chi)$ the same value as "Eisenstein summation".

This also proves that both the left-hand side and the right-hand side of (23) are independent of the choice of the generators u, v for W, provided $\chi(u) \neq 1$; by continuity, the same remains true for $\chi(u) = 1$, provided $\chi \neq 1$. For the right-hand side, this fact can also be verified directly by using formulas (15), (17) and

(25) of Chap. IV. A third proof, based on the expression of $X_q(z_0)X_{\bar{q}}(\bar{z}_0)$ by a double theta series depending only upon the quadratic form F, will be found in Kronecker (*Werke* IV, pp. 354—355); cf. Siegel's Bombay lectures (loc. cit. Chap. VII, § 13), pp. 54—57.

§ 12. For the analytic continuation of all series of type (1), two methods are available, both derived from the work of Kronecker. The first one is the one we have applied above to the series $G(s,\chi)$; it will therefore be enough to sketch it briefly in the general case.

Notations being as in § 2, and the constant A being defined by (15), § 9, we have

(26) $\qquad \chi(w) = \exp[A^{-1}(\bar{x}_0 w - x_0 \bar{w})]$.

We may now write, for the function defined by the series (1) for $\operatorname{Re}(s) > \dfrac{a}{2} + 1$:

(27) $\qquad K_a(x, x_0, s) = \sum{}^{*} \chi(w)(\bar{x}+\bar{w})^a |x+w|^{-2s}$,

where $\sum{}^{*}$ means the sum taken over all $w \in W$ other than $-x$ if x is in W; this depends only upon x, x_0, s and the lattice W. When it is necessary to be more explicit, we write for it $K_a(x, x_0, s; W)$. As always, a is an integer ≥ 0.

With the notations of Chap. VII, §§ 7 and 11, we have

(28) $\qquad K_a(x, x_0, s) = \sum_v \bar{u}^a |u|^{-2s} S_a(\zeta + v\tau, \delta\beta_0, s) e(\alpha_0 v)$.

Putting again $x = \alpha u + \beta v$, with α, β real, we have $\zeta = \alpha + \delta\beta\tau$. For $\zeta + v\tau$ to be real, we must have $v = -\delta\beta$. If β is an integer, we will write $A(s)$ for the term, in the right-hand side of (28), which corresponds to $v = -\delta\beta$; this is

$$A(s) = \bar{u}^a |u|^{-2s} S_a(\alpha, \delta\beta_0, s) e(-\delta\beta\alpha_0)$$

$$= \bar{u}^a |u|^{-2s} S_b\left(\alpha, \delta\beta_0, s - \frac{a-b}{2}\right) e(-\delta\beta\alpha_0),$$

where we have taken $b=0$ or 1, $b \equiv a \bmod 2$. If β is not an integer, we put $A(s) = 0$.

All the other terms in the right-hand side of (28) can be expressed by means of (39), Chap. VII, § 12; the latter formula gives for each such term a series, all of whose terms are given by (41) in that §, with the possible exception of one term given by (40), which occurs if and only if β_0 is an integer. If β_0 is an integer, collect all the terms thus given by (40), and call their sum $B(s)$; this gives:

$$i^a \bar{u}^{-a} |u|^{2s} \omega^{2s-a-1} B(s) = 2^{a+2-2s} \pi \frac{\Gamma(2s-a-1)}{\Gamma(s)\Gamma(s-a)} S_a\left(\delta\beta, -\alpha_0, s - \frac{1}{2}\right)$$

$$= \sqrt{\pi} \frac{\Gamma\left(s - \dfrac{a+b}{2}\right)}{\Gamma(s)\Gamma(s-a)} \Gamma\left(s - \frac{a+1-b}{2}\right) S_b\left(\delta\beta, -\alpha_0, s - \frac{a+1-b}{2}\right),$$

with b as above. If β_0 is not an integer, we put $B(s) = 0$.

All the remaining terms, which are those given by (41) of Chap. VII, § 12, make up a double series which we call $C(s)$; they all contain the Bessel function K_z, and our usual estimates for this show at once that $C(s)$ is absolutely convergent for all s and represents an entire function of s. As to $A(s)$ and $B(s)$, if they are not 0, they can be expressed by means of (14), Chap. VII, § 7; they are meromorphic functions in the whole s-plane, with at most simple poles, if $b=0$, at $s = \frac{a+1}{2}$ and $s = \frac{a}{2} + 1$. Actually the calculation of the residues at those points, as given by (14) of Chap. VII, § 7, shows that at most $s=1$ can be a pole, and this only if $a=0$, $\chi=1$.

§ 13. The same results can be obtained, even more naturally perhaps, by the second method, which is the same as the one applied in Chap. VII, § 7, to the analytic continuation of the simple series $S_a(x,y,s)$ for real values of x.

Applying Dirichlet's transformation to (27), we get:

$$(29) \qquad \Gamma(s) K_a(x, x_0, s) = \int_0^{+\infty} \Theta_a^*(t, x, x_0) t^{s-1} dt,$$

where we have put:

$$\Theta_a^*(t, x, x_0) = \sum{}^* \exp[-t|x+w|^2] \chi(w)(\bar{x}+\bar{w})^a$$
$$= \sum{}^* \exp[-t|x+w|^2 + A^{-1}(\bar{x}_0 w - x_0 \bar{w})](\bar{x}+\bar{w})^a.$$

Here, as before, $\sum{}^*$ means the sum extended to all $w \in W$ other than $-x$. We will write Θ_a for the similar sum extended to all $w \in W$; then we have $\Theta_a^* = \Theta_a - \chi(-x)$ if $a=0$ and $x \in W$, and $\Theta_a^* = \Theta_a$ in all other cases.

Extend χ to a character of the additive group of $\mathbf{C}$ by putting, for all $x \in \mathbf{C}$:

$$\chi(x) = \exp[A^{-1}(\bar{x}_0 x - x_0 \bar{x})];$$

then the function $\Theta_a(t, x, x_0) \chi(x)$ is periodic in x, with the period-lattice W; if we write again $x = \alpha u + \beta v$, it is periodic in α, β with the period 1; expanding it into a Fourier series in α, β, and using the Fourier formulas to calculate the coefficients, amounts to doing Poisson summation on Θ_a. It is even more convenient to treat first Θ_0 in this manner, and then obtain the formula for Θ_a by applying $\partial^a / \partial x^a$ to both sides (cf. the similar procedure for simple theta-series in Chap. VII, § 7). One gets

$$(30) \qquad \Theta_a(t, x, x_0) = (At)^{-a-1} \Theta_a(A^{-2} t^{-1}, x_0, x) \chi(-x).$$

Now we cut up the integral in (29) into one over the interval $0 < t \leq T$ and one over $t > T$; the latter, for which we write $I_\infty(T, x, x_0, a, s)$ is absolutely convergent for all s and defines an entire function of s. In the other one, we express Θ_a^* in terms of Θ_a, then apply (30), and finally make the change of variable $t = A^{-2}u^{-1}$. This gives:

$$
\begin{aligned}
(31)\quad \Gamma(s)K_a(x,x_0,s) &= I_\infty(T,x,x_0,a,s) \\
&\quad + A^{a+1-2s} I_\infty(A^{-2}T^{-1},x_0,x,a,a+1-s)\chi(-x) \\
&\quad - \varepsilon \frac{T^s}{s}\chi(-x) + \varepsilon' A^{-a-1} \frac{T^{s-a-1}}{s-a-1},
\end{aligned}
$$

where $\varepsilon = 1$ if $a = 0$, $x \in W$, and $\varepsilon = 0$ otherwise, and $\varepsilon' = 1$ if $a = 0$, $x_0 \in W$, and $\varepsilon' = 0$ otherwise.

This gives the analytic continuation of the left-hand side as a meromorphic function in the whole s-plane, with possible poles only at $s=0$ (if $a=0$, $x \in W$) and at $s=1$ (if $a=0$, $x_0 \in W$). If $a=0$, $x \in W$, the residue at $s=0$ is $-\chi(-x)$; it is -1 if $x=0$. If $a=0$, $x_0=0$ (i.e. $\chi=1$), the residue at $s=1$ is A^{-1}.

Taking $T=A^{-1}$ in (31), we get the functional equation

$$
(32)\quad \Gamma(s)K_a(x,x_0,s) = A^{a+1-2s}\Gamma(a+1-s)K_a(x_0,x,a+1-s)\chi(-x).
$$

We also note that, for $\operatorname{Re}(s) > \frac{a}{2}+1$, $K_a(x,0,s)$ is an even or an odd function of x, according as a is even or odd, and that it is periodic in x with the period-lattice W; therefore, by analytic continuation, the same is true for all s.

§ 14. We are now in a position to settle the question which had to be left open at the end of Chap. VI. Consider again the functions $E_{a,b}$, $E_{a,b}^*$ defined by (8) of Chap. VI, § 4. Obviously we have, for x not in W:

$$
E_{a,b}(x) = E_{a,b}^*(x) = K_{a+b}(x,0,b)
$$

provided the series defining $E_{a,b}$ is absolutely convergent, i.e. for $b \geq a+3$. It will now be shown that we still have

$$
(33)\quad E_{a,b}^*(x) = K_{a+b}(x,0,b)
$$

for $b-a=1$ or 2. In fact, we have, for $\operatorname{Re}(s) > \frac{a}{2}+1$:

$$
(34)\quad \frac{\partial}{\partial x} K_a(x,x_0,s) = -s K_{a+1}(x,x_0,s+1).
$$

Similarly, we have, if $a > 0$:

$$
(35)\quad \frac{\partial}{\partial \bar{x}} K_a(x,x_0,s) = (a-s) K_{a-1}(x,x_0,s).
$$

If $\mathscr{D}$ is the differential operator defined by (1), Chap. VI, § 1, we have also:

(36) $$\mathscr{D}K_a(x,0,s) = -sK_{a+2}(x,0,s+1).$$

By analytic continuation, these formulas remain valid for all s.

In particular, (35) shows that $K_a(x,x_0,a)$ is a holomorphic function of x, for $a>0$ and x not in W, except when $K_{a-1}(x,x_0,s)$ has a pole at $s=a$; as we have seen in § 13, this can happen only if $a=1$ and $x_0 \in W$. As we have found there that the residue of $K_0(x,0,s)$ at $s=1$ is A^{-1}, this gives:

$$\frac{\partial}{\partial \bar{x}} K_1(x,0,1) = -A^{-1}.$$

Now consider first the case of $K_2(x,0,2)$; the above formulas show that it has the same derivatives, with respect to x and to $\bar{x}$, as the function $E_{0,2}$, i.e. as E_2; therefore it differs from E_2 only by a constant. Now consider $K_1(x,0,1) - E_1(x)$; its derivatives with respect to x and to $\bar{x}$ are constant, and it is an odd function of x; therefore it is a real-linear function of x. As $K_1(x,0,1)$ is periodic in x with the period-lattice W, this, as we have seen in Chap. VI, § 2, implies that it is no other than E_1^*. Now, using (34) and (36), together with (8) of Chap. VI, § 4, we get (33) for all $b > a \geq 0$.

§ 15. It follows at once from (27) that we have, at first for $\mathrm{Re}(s) > \frac{a}{2} + 1$ and then, by analytic continuation, for all s:

$$K_a(0,x_0,s) = [K_a(x,x_0,s) - \bar{x}^a|x|^{-2s}]_{x=0}.$$

In view of the definition of $e^*_{a,b}$ in Chap. VI, § 5, one concludes at once from this, and from the results in § 14, that we have

$$e^*_{a,b} = K_{a+b}(0,0,b)$$

whenever $b > a \geq 0$. With this, we have fully justified all the assertions made at the end of Chap. VI. In other words, writing now $N = a + b$, we have proved that

$$\pi^{N-b} \varpi^{-N} K_N(x,0,b)$$

is algebraic over $\mathbf{Q}$ whenever $\mathbf{Q}W$ is an imaginary quadratic field k, $x \in k$, and $b \leq N$, $N < 2b$.

To complete the matter, so far as we are concerned here, consider the case where the character χ entering into the definition of $K_a(x,x_0,s)$ is of finite order; this is so if and only if x_0 is in $\mathbf{Q}W$, or also if and only if the kernel of the character χ of W is a sublattice W' of W. In that case, take a set R of representatives of W/W' in W. Then we have, at first for $\mathrm{Re}(s) > \frac{a}{2} + 1$ and then, by analytic continuation, for all s:

$$K_a(x, x_0, s; W) = \sum_{r \in R} \chi(r) K_a(x+r, 0, s; W').$$

This, combined with the result stated above, shows that

$$\pi^{N-b} \varpi^{-N} K_N(x, x_0, b)$$

is algebraic over $\mathbf{Q}$ whenever $W \subset k$, $x \in k$, $x_0 \in k$, $0 < N/2 < b \leq N$.

Finally, if we make use of the functional equation (32), we conclude that this last statement remains true for $0 < b \leq N$. We leave it to the reader to translate this into a similar statement concerning the values of Hecke L-functions over imaginary quadratic fields.

§ 16. In Chap. VII, § 10, it has been shown how some of the results of that chapter could be re-interpreted in the light of the theory of distributions. The same will be done now for Kronecker's double series.

In the complex plane $\mathbf{C}$, we take $|dx\,d\bar{x}|$ as the normalized Haar measure. If W is a lattice in $\mathbf{C}$, $T = \mathbf{C}/W$ is a complex torus; if we put $x = \alpha u + \beta v$ as usual, we have $|dx\,d\bar{x}| = 2\pi A\,d\alpha\,d\beta$, and $d\alpha\,d\beta$ is the Haar measure on T, normalized by $\int_T d\alpha\,d\beta = 1$. If f is an integrable function on T, then

(37) $$\varphi \mapsto \Delta(\varphi) = \int_T \varphi f\,d\alpha\,d\beta$$

is a distribution on T, for which we write $\Delta = f\,d\alpha\,d\beta$, or, for the sake of brevity, simply f. By "abuse of language", we say then that Δ "is" the function f.

Fourier series on $T = \mathbf{C}/W$ proceed according to the characters of T, or, what amounts to the same, to the characters of $\mathbf{C}$ which are trivial on W. These can be written:

(38) $$x = \alpha u + \beta v \mapsto \psi_{\mu\nu}(x) = \mathbf{e}(\delta v \alpha - \delta \mu \beta) = \exp[A^{-1}(w\bar{x} - \bar{w}x)],$$

where $w \in W$, $w = \mu u + \nu v$, as usual. Then, if Δ is any distribution on T, we associate with it the Fourier series

$$\sum_{\mu,\nu} \Delta(\psi_{\mu\nu}^{-1}) \psi_{\mu\nu}.$$

Conversely, it is well-known that any formal Fourier series on T determines a distribution, provided its coefficients are $O((\mu^2 + \nu^2)^N)$ for some N.

We say (cf. Chap. VII, § 10) that a distribution in the complex plane $\mathbf{C}$ "belongs" to a quasicharacter ω of the multiplicative group $\mathbf{C}^\times$ if it gets multiplied by ω under the action of that group. There is, up to a constant factor, one and only one such distribution, once ω is given. If $\omega(x) = x^{-m}\bar{x}^{-n}$, where m, n are integers ≥ 0, the distribution $(\partial^{m+n}/\partial x^m \partial \bar{x}^n)_{x=0}$ belongs to ω; it has the support 0.

The quasicharacters of $\mathbf{C}^\times$ are the functions of the form

$$x \mapsto \omega(x) = \bar{x}^a |x|^{2-2s},$$

where $s \in \mathbf{C}$ and a is an integer ≥ 0, or their complex-conjugates. Unless this is of the form $x^{-m}\bar{x}^{-n}$ with integers $m, n \geq 0$, a distribution belonging to ω is given by the formula

$$D_\omega(\Phi) = \mathrm{Pf} \int \Phi(x)\omega(x)|x|^{-2}|dx\,d\bar{x}|\,;$$

by definition, this means the limit of

$$\int_{x\bar{x} \geq \varepsilon} \Phi(x)\omega(x)|x|^{-2}|dx\,d\bar{x}| - \varepsilon^{a+1-s}P(\varepsilon)$$

for $\varepsilon \to 0$ when P is a polynomial so chosen that this has a finite limit. The right-hand side is well-defined provided: (a) Φ is everywhere locally integrable; (b) it is m times continuously differentiable near $x=0$, with $m > 2\mathrm{Re}(s) - a - 2$; (c) for $|x| \to \infty$, it is $O(|x|^\lambda)$ with $\lambda < 2\mathrm{Re}(s) - a - 2$.

From now on, assume that $\mathrm{Re}(s) > \dfrac{a}{2} + 1$; then the above remark implies that $D_\omega(\Phi)$ is well defined if we take for Φ any character of the additive group of $\mathbf{C}$. Write ψ for the character $\psi(x) = \mathbf{e}(x+\bar{x})$, and ψ_b for $\psi_b(x) = \psi(bx)$, so that $\psi_0 = 1$. Put $\hat{B}_\omega = D_\omega(\psi)$ and $f(b) = D_\omega(\psi_b)$. As D_ω belongs to ω, we have $f(by) = f(b)\omega(y)^{-1}$ for all b and all $y \neq 0$. As $\omega \neq 1$, this gives $f(0) = 0$, i.e. $D_\omega(1) = 0$, and, for $b \neq 0$, $f(b) = B_\omega \cdot \omega(b)^{-1}$.

There are various ways of calculating the constant B_ω; as this belongs more properly to the theory of the "Tate distribution" D_ω and of its Fourier transform, we merely give the result:

$$B_\omega = i^a (2\pi)^{2s-a-1} \frac{\Gamma(a+1-s)}{\Gamma(s)}.$$

§ 17. Now, with our usual notations, we consider again the lattice W and the torus $T = \mathbf{C}/W$. For a given $x_0 \in \mathbf{C}$, we introduce once more the character χ of $\mathbf{C}$ given by

$$\chi(x) = \exp[A^{-1}(\bar{x}_0 x - x_0 \bar{x})]$$

which we could also write now as ψ_b with $b = (2\pi i A)^{-1}\bar{x}_0$. Let φ be a bounded integrable function on the torus T, indefinitely differentiable near 0. We attach to it the function

$$\Phi(x) = \varphi(x \bmod W)\chi(x)$$

on $\mathbf{C}$, and write Δ, or, when necessary, $\Delta(\omega, x_0)$ for the distribution on T given by $\Delta(\varphi) = D_\omega(\Phi)$, this being well-defined since we have assumed $\operatorname{Re}(s) > \frac{a}{2} + 1$. In an obvious sense, we can write:

$$\Delta(\varphi) = \operatorname{Pf} \int_T \varphi(x) K_a(x, x_0, s) \chi(x) |dx \, d\bar{x}|$$

where the integrand, being periodic in x with the period-lattice W, should be regarded as a function on T.

If the support of φ is disjoint from 0, the symbol Pf becomes unnecessary. In the sense explained above, one may say that, outside 0, the distribution Δ "is" the function

(39) $\qquad 2\pi A K_a(x, x_0, s) \chi(x).$

It is now easy to write the Fourier series for Δ; according to our definitions, the coefficient, in that series, of the character $\psi_{\mu\nu}$ given by (38) has the value $D_\omega(\Phi)$ when Φ is the function

$$\Phi(x) = \exp[A^{-1}(\bar{x}_0 + \bar{w})x - A^{-1}(x_0 + w)\bar{x}].$$

As Φ is no other than the character ψ_b of $\mathbf{C}$ with b given by

$$b = (2\pi i A)^{-1}(\bar{x}_0 + \bar{w}),$$

the coefficient in question has the value 0 if $b = 0$, and otherwise the value $f(b) = B_\omega \cdot \omega(b)^{-1}$. Thus the Fourier series for Δ is:

(40)
$$2\pi A^{a+2-2s} \frac{\Gamma(a+1-s)}{\Gamma(s)} \sum{}^* (\bar{x}_0 + \bar{w})^a |x_0 + w|^{2(s-a-1)} \cdot \exp[A^{-1}(w\bar{x} - \bar{w}x)]$$

$$= 2\pi A^{a+2-2s} \frac{\Gamma(a+1-s)}{\Gamma(s)} K_a(x_0, x, a+1-s),$$

the last factor in the right-hand side being understood formally, since it need not converge.

Writing this to be "equal" to (39), we get a formula which is formally identical with the functional equation (32), but of course with a different significance.

§ 18. When $\operatorname{Re}(s) \leq \dfrac{a}{2} + 1$, the Fourier series (40) still defines a distribution, provided s is not a pole of $\Gamma(a+1-s)$, i.e., provided ω is not of the form $x^{-m}\bar{x}^{-n}$ with integers $m, n \geq 0$. The distribution thus defined by (40) may be regarded, in an obvious sense, as deduced from $\Delta(\omega, x_0)$ by analytic continuation in the s-plane. Analytic continuation shows that, outside 0, it is still "equal" to the function (39). Actually, for $\operatorname{Re}(s) < a/2$, (40) is absolutely convergent, so that its sum is equal to (39) in the usual sense.

We will not pursue this any further, but conclude with an interesting example. Consider the distribution Δ_0 given on T by the Fourier series

$$K_0(0, x, 1) = \sum' |w|^{-2} \exp[A^{-1}(w\bar{x} - \bar{w}x)] \; ;$$

except for the substitution of x for x_0, this is the same as the series $G(1, \chi)$ of §§ 9—10, for which we proved Kronecker's formula, i.e. (23) of § 9. We shall now obtain the same result by a different method.

We will write $\mathscr{L}$ for the "Laplacian" (or "Beltrami operator") $\partial^2/\partial x \partial \bar{x}$ on T. As the Fourier series of distributions may be differentiated term by term, we have

(41) $\qquad \mathscr{L}(\Delta_0) = A^{-2}\left(1 - \sum_{w \in W} \exp[A^{-1}(w\bar{x} - \bar{w}x)]\right).$

The series, which is the formal sum of all the characters of T, is the Fourier series for the "Dirac distribution" μ_0, i.e. the mass 1 at 0; accordingly, the right-hand side of (41) is $A^{-2}(1 - \mu_0)$. Any two solutions of this equation must differ by a harmonic function on T, hence by a constant; therefore (41), together with the condition $\Delta_0(1) = 0$, determines Δ_0 uniquely; in particular, these conditions imply that Δ_0 is real.

If we put, as before (cf. Chap. VI, § 2) $\beta = \beta(x)$ when $x = \alpha u + \beta v$, a trivial calculation gives

$$\mathscr{L}[\beta(x)^2] = \tfrac{1}{2} |u/\pi A|^2.$$

On the other hand, it is well-known that one has

$$\mathscr{L}[\log(x\bar{x})|dx\,d\bar{x}|] = 2\pi \mu_0$$

in the sense of distribution-theory, this being an almost immediate consequence of the definitions; this gives

$$\mathscr{L}[\log(x\bar{x})d\alpha\,d\beta] = A^{-1}\mu_0.$$

Finally, it is well-known that a real-valued distribution is harmonic if and only if it is everywhere locally of the form $\log|f(x)|^2$, with f holomorphic and $\neq 0$.

Now consider the function

$$F(x) = 2\pi^2 |u|^{-2} \beta(x)^2 - A^{-1} \log |X_q[e(x/u)]|^2 ;$$

as we have already noted in §9, this is periodic in x with the period-lattice W and may be regarded as a function on the torus T. From the above observations it follows that $\Delta_0 - F(x)$ is a harmonic distribution on T and therefore a constant. Since $\log|x|^2$ is locally integrable near $x=0$, F is integrable on the torus; as $\Delta_0(1) = 0$, we get:

$$\Delta_0 = F - \int_T F \, d\alpha \, d\beta .$$

The calculation of the last integral is a routine exercise. The conclusion is, as expected, Kronecker's formula

$$\Delta_0 = F + \tfrac{1}{3} \pi^2 |u|^{-2}$$

which shows at the same time that Δ_0 is "equal" to an integrable function on T. It is no other than Green's function on the torus T for the Riemannian metric $ds^2 = |dx|^2$.

Chapter IX
Finale: Allegro con brio

(Pell's equation and the Chowla-Selberg formula)

§ 1. Eisenstein, having laid the foundations for a theory of elliptic functions, was able to carry out much of his design for the building itself, and to indicate how he wished it completed. Kronecker, having conceived ambitious plans for a vastly enlarged edifice, started, rather late in life, to dig deeper foundations but found time for little else. It is idle to speculate about the kind of continuation he had in mind; perhaps he did not know it himself.

On the other hand, the number-theoretic motivation for the huge efforts of his later years can be discerned clearly and deserves a short digression here. Kummer, then Dirichlet, had been his teachers and remained his lifelong friends. Kummer's arithmetical work centered around cyclotomic fields, their ideal-classes and their class-numbers. As Dirichlet had first discovered in a different setting, those class-numbers depend upon the values of Dirichlet's L-functions at $s=1$, i.e. ultimately upon the values, for suitable values of the arguments, of the simple series discussed above in our Chap. VII. Kummer, having re-established those results in the language of his ideal-theory, proceeded to investigate the p-adic properties of those values, beginning with his celebrated congruences for Bernoulli numbers. This had in fact been the deepest part of his work, more valuable perhaps than the more spectacular applications to Fermat's problem and even to the reciprocity laws.

Early in life (in 1853), Kronecker found the conceptual justification for Kummer's singling out of the cyclotomic fields; not only are they (as Gauss had discovered) abelian extensions of the rational number-field, but they are the only ones.

Already in the same paper (*Werke* IV, pp. 10—11) where he announced this momentous discovery, Kronecker stated the corresponding result for the Gaussian field $\mathbf{Q}(i)$ and the lemniscatic functions. Out of this grew the idea that the division of elliptic functions with complex multiplication must play the same role, for the corresponding imaginary quadratic fields, as the division of the circle plays for $\mathbf{Q}$, and that of the lemniscate for $\mathbf{Q}(i)$. Allusions to this occur in the letter to

Dirichlet of May 1857 (*Werke* V, p. 420; cf. also *Werke* IV, pp. 179—183). This, as he wrote later to Dedekind, had been the most cherished dream of his youth ("*mein liebster Jugendtraum*"; *Werke* V, p. 455).

Frequently this has been taken to mean no more than the extension to imaginary quadratic fields of his existence theorem of 1853 for $\mathbf{Q}$, or in other words the conjecture that the division of elliptic functions with complex multiplication supplies all the abelian extensions of such fields. But this seems too restrictive an interpretation.

It is true that his memoir of 1886 (*Werke* IV, pp. 389—470), where he proves his famous congruences, must have been written chiefly with that conjecture in mind. He did not prove it there; but he supplied all the tools for doing so, leaving the completion of this task to his successors.

Nevertheless, the conjecture may well have been only part of the dream, perhaps no more than its beginning. It was the whole work of Kummer on the factors of the class-numbers of cyclotomic fields and on their p-adic properties that he wished to extend to all imaginary quadratic fields and their abelian extensions.

Such seems to have been the huge program which Kronecker was planning to carry out in his Berlin academy memoirs. Needless to say, he did not fulfil it.

It was only given to him to bring to some degree of completeness the analytical part. A whole galaxy of later writers, H. Weber, Fueter, Hasse, Hecke, C. Meyer, Siegel, K. Ramachandra, have taken it up since, perhaps without exhausting even the purely arithmetical problems[12]. As to the p-adic investigation, it has barely begun; Eisenstein's great paper of 1850 on the lemniscate, extending to that case so much of Kummer's work on cyclotomy, has been largely forgotten, and Hurwitz's and Herglotz's contributions have remained quite isolated until recently. As this is outside our subject, no more can be done here than to point this out as a promising topic for further study.

§ 2. Here we shall merely give two illustrations, both taken from Kronecker, for the kind of application he had in mind. The first one reproduces the content of his paper of 1863 on "Pell's equation" (*Werke* IV, pp. 221—225; cf. also *Werke* IV, pp. 379—389). The other one, as Kronecker left it (*Werke* IV, pp. 376—379), needed an additional ingredient to make it complete; this was supplied by Lerch in 1894 (it is the formula (23) of our Chap. VII, § 9); but the two were first put together by Chowla and Selberg in 1949 (*Proc. Nat. Ac. USA* 35, pp. 371—374; cf. also *Crelles J.* 227 (1967), pp. 86—110). In our exposition, we shall take for granted some basic results from class-field theory; needless to say, Dirichlet and Kronecker possessed direct proofs for them.

---

[12] For bibliographies and useful historical surveys, as well as for perhaps the most substantial results up to date, cf. C. Meyer, *Berechnung der Klassenzahl*... (Ak.-Verlag, Berlin 1957); C.L. Siegel, *Lectures on advanced analytic number-theory* (T.I.F.R., Bombay 1961), Chap. II; K. Ramachandra, *Some applications of Kronecker's limit-formulas*, Ann. of Math. 80 (1964), 104—148.

In 1841, Dirichlet had extended to the Gaussian field $\mathbf{Q}(i)$ much of his work on the L-series over $\mathbf{Q}$, observing at the same time that the summation of such series depends upon the lemniscatic functions (Dirichlet, *Werke* I, pp. 503—618); his theory of quadratic forms over $\mathbf{Q}(i)$ amounted to a study of quadratic extensions of $\mathbf{Q}(i)$, and he noted the remarkable properties of biquadratic fields of the form $\mathbf{Q}(i,\sqrt{m})$ and of the zeta-functions and L-functions associated with them (ib., p. 508 and pp. 612—618). Such was the background for Kronecker's paper of 1863.

Viewed from a modern vantage point, Kronecker's idea looks simple enough. Let K be a field with a Galois group of type $(2,2)$ over $\mathbf{Q}$; then it has three quadratic subfields k_0, k_1, k_2, and it is the compositum of any two of them. One assumes that K is not real; then one and only one of the fields k_α, say k_2, is real; let $\varepsilon > 1$ be a fundamental unit in k_2.

Let $\zeta, \zeta_\alpha, \zeta_K$ be the zeta-functions of $\mathbf{Q}, k_\alpha, K$, respectively; put $L_\alpha = \zeta_\alpha/\zeta$; this is the Dirichlet L-function over $\mathbf{Q}$, attached to the quadratic field k_α. Then $\zeta_K = \zeta L_0 L_1 L_2$. Putting $s=1$, one gets a relation between the class-numbers of K, k_0, k_1, k_2; in a different language, and restricted to the case $k_0 = \mathbf{Q}(i)$, this had been one of Dirichlet's main results.

Now put $\Lambda = \zeta_K/\zeta_0 = L_1 L_2$. In this identity, which had also been considered by Dirichlet for $k_0 = \mathbf{Q}(i)$, Λ is the L-series over k_0 attached to the quadratic extension K of k_0; it is therefore a linear combination of double series of the type discussed in our Chap. VIII, while L_1, L_2 are similar combinations of the simple series of our Chap. VII; thus $\Lambda = L_1 L_2$ is a non-linear identity of arithmetical origin between such series. In it, put $s=1$; then the right-hand side can be expressed by Dirichlet's formulas in terms of the class-numbers for k_1, k_2 and of the unit ε in k_2. As to $\Lambda(1)$, it can be expressed in terms of elliptic functions by means of Kronecker's limit-formula; the functions which occur here are those with complex multipliers from the field k_0. Thus we have a unit of k_2, i.e. a solution of "Pell's equation", expressed in terms of elliptic functions.

This was Kronecker's discovery of 1863. It was sensational enough at the time.

§ 3. Kronecker's point of view is to regard k_0 as given and K as the simplest case of an abelian extension of k_0, chosen so as to be abelian also over $\mathbf{Q}$. Actually, in order to have more manageable formulas, Kronecker only worked out the case where K is unramified over k_0; we shall also make that assumption here, for the same reason[13].

For $\alpha = 0, 1, 2$, call D_α the discriminant of k_α, h_α its class-number and w_α the number of roots of unity in k_α; as k_2 is real, w_2 is 2. Put $m_\alpha = |D_\alpha|$; then $D_0 = -m_0$, $D_1 = -m_1, D_2 = m_2$. The Dirichlet character attached to k_α is given by the Legendre

---

[13] Kronecker warns that similar results are valid without that assumption; this is the purport of his words "**und** selbst dann, wenn dieselben einen gemeinsamen Factor haben", *Werke* IV, p. 223.

symbol $\chi_\alpha(n) = (D_\alpha/n)$. Then we have:

(1) $$L_\alpha(s) = \sum_{n=1}^{+\infty} \chi_\alpha(n) n^{-s} = m_\alpha^{-s} \sum_{n=1}^{m_\alpha - 1} \chi_\alpha(n) H\left(\frac{n}{m_\alpha}, s\right),$$

with $H(x,s)$ as defined in Chap. VII, § 9. The value of L_α at $s=1$ is given either by the results in our Chap. VII or (what amounts to the same) by Dirichlet's classical formulas. Of course Kronecker followed the latter procedure, which gives

(2) $$L_1(1) = \frac{2\pi\, h_1}{w_1 \sqrt{m_1}}, \qquad L_2(1) = \frac{2h_2}{\sqrt{m_2}} \log \varepsilon.$$

On the other hand, it is well-known that K is unramified over k_0 if and only if $D_0 = D_1 D_2$ and D_1, D_2 are mutually prime. When that is so, the character χ attached to K over k_0 is a so-called "genus character", i.e. a character of order 2 of the group of ideal-classes of k_0. Therefore, in the series

(3) $$\Lambda(s) = \sum \chi(\mathfrak{m}) N(\mathfrak{m})^{-s},$$

all the ideals belonging to the same class have the same coefficient $\chi(\mathfrak{m})$.

Let $\mathfrak{a}$ be any fractional ideal in k_0. In (3), collect together all the ideals equivalent to $\mathfrak{a}^{-1}$; they are those of the form $\alpha \mathfrak{a}^{-1}$ with $\alpha \in \mathfrak{a}$, $\alpha \neq 0$; $\alpha' \mathfrak{a}^{-1}$ is the same as $\alpha \mathfrak{a}^{-1}$ if and only if α'/α is a root of unity. Therefore, collecting the corresponding terms in (3) together, we get the series

(4) $$w_0^{-1} \chi(\mathfrak{a})^{-1} N(\mathfrak{a})^s \sum_{\alpha \in \mathfrak{a}}{}' |\alpha|^{-2s};$$

here w_0 must be 2, since otherwise D_0 would be -3 or -4 and could not be written as $D_1 D_2$ with D_1, D_2 as above.

We may regard $\mathfrak{a}$ as a lattice in $\mathbf{C}$ and apply to it Kronecker's limit-formula, i.e. (16) of Chap. VIII, § 8; here we have to put $W = \mathfrak{a}$, so that the constant A in that formula, as defined by (15) in the same §, has the value

$$A = \frac{N(\mathfrak{a})}{2\pi} \sqrt{m_0}.$$

We have $\bar{\mathfrak{a}} = N(\mathfrak{a})\mathfrak{a}^{-1}$. Put $\Delta = \Delta(\mathfrak{a})$, with $\Delta(W)$ as defined in Chap. IV, § 11; since it is homogeneous of degree -12 in u, v, i.e. in W, we have

$$\bar{\Delta} = \Delta(\bar{\mathfrak{a}}) = N(\mathfrak{a})^{-12} \Delta(\mathfrak{a}^{-1}).$$

Accordingly, we will write

$$F(\mathfrak{a}) = \Delta(\mathfrak{a})\Delta(\mathfrak{a}^{-1}) = N(\mathfrak{a})^{12}|\Delta(\mathfrak{a})|^2;$$

clearly this depends only upon the ideal-class of $\mathfrak{a}$.

Applying now Kronecker's formula to (4), we get

(5) $$N(\mathfrak{a})^s \sum |\alpha|^{-2s} = \frac{2\pi}{\sqrt{m_0}}\left(\frac{1}{s-1} + C_0 - \frac{1}{12}\log F(\mathfrak{a}) + \cdots\right),$$

where C_0 is a sum of terms depending only upon k_0, not upon $\mathfrak{a}$.

Take representatives $\mathfrak{a}_i$ for the ideal-classes of k_0. As χ is of order 2, we have $\chi = \chi^{-1}$; as it is not trivial, we have $\sum \chi(\mathfrak{a}_i) = 0$. Combining (4) and (5), we get

$$\Lambda(1) = -\frac{1}{12}\frac{\pi}{\sqrt{m_0}}\sum \chi(\mathfrak{a}_i)\log F(\mathfrak{a}_i);$$

this, combined with (2) and with $\Lambda = L_1 L_2$, gives Kronecker's final result:

(6) $$h_1 h_2 \log \varepsilon = -\frac{w_1}{48}\sum \chi(\mathfrak{a}_i) \log F(\mathfrak{a}_i).$$

Instead of putting $s=1$ in $\Lambda = L_1 L_2$, we could have put $s=0$, and applied formula (17) of Chap. VIII, § 8, to the left-hand side, and (21) of Chap. VII, § 9, to the right-hand side. In that way, we would have had the right-hand side of (6) expressed in an "elementary way" in terms of the "cyclotomic unit" ε^{h_2}. Of course the two results are equivalent.

§ 4. As Kronecker observed, one can proceed exactly in the same way with the identity $\zeta_0 = \zeta L_0$. Here we have

$$\zeta_0(s) = \sum N(\mathfrak{m})^{-s} = \frac{1}{w_0}\sum_i N(\mathfrak{a}_i)^s \sum_{\alpha \in \mathfrak{a}_i}{}' |\alpha|^{-2s},$$

and Kronecker's limit-formula gives the first two terms of the expansion of ζ_0 at $s=1$. It amounts to the same to consider the expansion at $s=0$, but the calculations are simpler. Formula (17) of Chap. VIII, § 8, shows, firstly, that $\zeta_0(0) = -h_0/w_0$; this agrees of course with the fact that $\zeta(0) = -\frac{1}{2}$ and $L_0(0) = 2h_0/w_0$. Secondly, we get:

$$\zeta_0'(0) = -\frac{1}{12 w_0}\sum \log F(\mathfrak{a}_i).$$

Instead of this, Kronecker obtains the equivalent formula for the first two terms of the expansion of ζ_0 at $s=1$. This being done, however, he merely expresses this in terms of known quantities and of the series

$$L'_0(1) = - \sum_{n=1}^{+\infty} \chi_0(n) \frac{\log n}{n}.$$

To get a complete result, one must make use of Lerch's theorem (i.e. (23) of Chap. VII, §9), or, as Chowla and Selberg do (loc. cit.), some equivalent result, e.g. Kummer's series for $\log \Gamma(s)$. Lerch's theorem, together with (1), gives:

$$L'_0(0) = -L_0(0) \log m_0 + \sum_{n=1}^{m_0-1} \chi_0(n) \log \Gamma(n/m_0).$$

Equating the derivatives of ζ_0 and of ζL_0 at $s=0$, one obtains the Chowla-Selberg formula. In writing it, we shall now write k, m, h, w, χ instead of $k_0, m_0, h_0, w_0, \chi_0$; then the formula is (writing numbers instead of their logarithms)

(7) $$\prod_{i=1}^{h} F(\mathfrak{a}_i) = \left(\frac{2\pi}{m}\right)^{12h} \prod_{n=1}^{m-1} \Gamma\left(\frac{n}{m}\right)^{6w\chi(n)}.$$

For instance, if $h=1$, we can take for $\mathfrak{a}_1$ the ring Ω of all integers in k. Then, with the notations of Chap. V, §8, the left-hand side of (7) is ϖ^{24}; so the constant ϖ is expressed in terms of values of $\Gamma(s)$ for rational arguments. In general, we know, from the definition of Δ in Chap. IV, §11, and the final results of Chap. VI, §6, that, for all $\mathfrak{a}$, $\varpi^{-12} \Delta(\mathfrak{a})$ is algebraic over $\mathbf{Q}$. Consequently, ϖ differs from the quantity

$$\sqrt{\pi} \prod_{n=1}^{m-1} \Gamma\left(\frac{n}{m}\right)^{w\chi(n)/4h}$$

only by a factor which is algebraic over $\mathbf{Q}$. The same is therefore true of the periods of any elliptic function-field with a complex multiplier in k, provided this field is defined by an equation with algebraic coefficients over $\mathbf{Q}$.

Index of Notations

$\exp(x) = e^x$; $\mathbf{e}(x) = \exp(2\pi i x)$.

Chapter II: $\varepsilon_n(x)$, $\sum_{e}^{\mu}$ ("simple Eisenstein summation"), $\sum_e$, §1, p. 6; γ_m, §1, p. 7; $\prod_e$, $\prod_e^{\mu}$, §6, p. 11; B_m, §7, p. 13.

Chapter III: $W, u, v, \delta, \tau, q, A$, §1, p. 14; $E_n(x)$, $\sum_{e}^{\mu,\nu}$ ("double Eisenstein summation"), $\sum_e$, §2, p. 14; W', u', v', (a,b,c,d), R, §5, p. 17; ζ, z, §6, p. 18; e_m, §7, p. 20; R', §8, p. 21.

Chapter IV: $f(t,x)$, $\varphi(x)$, §3, p. 25; $X_q(z)$, $P(q)$, §3, p. 26; $T(q,z)$, §6, p. 28; $\eta(\tau)$, §8, p. 29; $\theta(\zeta,\tau)$, §8, p. 30; Δ, $\Delta(u,v)$, $\Delta(W)$, §11, p. 34.

Chapter V: $e(W)$, $E(W)$, e, E, §2, p. 37; ω ("complex mutiplier"), §6, p. 39; ϖ, ϖ_m, §8, p. 41.

Chapter VI: $\mathcal{D}$, §1, p. 42; A, α, β, $\beta(x)$, E_n^*, e_2^*, §2, p. 43; $E_{a,b}$, $E_{a,b}^*$, §4, p. 44; $e_{a,b}$, $e_{a,b}^*$, §5, p. 45.

Chapter VII: $\chi(\mu)$, $S_a(x,y,s)$, $\sum^*$, §4, p. 53; $S_a(x,y,s)$ (for $a=0$ or 1, $x \in \mathbf{R}$), §7, p. 56; $H(x,s)$, §9, p. 58; $S_a(\zeta,y,s)$ (for $a \geq 0$, $\zeta \in \mathbf{C}$, $\zeta \notin \mathbf{R}$), §11, p. 63; K_z, §12, p. 65.

Chapter VIII: $\chi(w)$, $\chi(\mu,\nu)$, §1, p. 69; ζ, τ, z, q, α_0, β_0, x_0, ζ_0, z_0, $F(q,z,w)$, §2, p. 70; $G(s,\chi)$, §6, p. 72; $\tau = \xi + i\omega$, §7, p. 73; A, §8, p. 74; $K_a(x,x_0,s)$, §12, p. 78; Θ_a^*, Θ_a, §13, p. 79.

Printing and binding: Druckerei Triltsch, Würzburg